# Molecular and Cellular Technologies for Forage Improvement

# Related Society Publications

*Alfalfa and Alfalfa Improvement*

*Alfalfa Management Guide, Revised*

*Clover Science and Technology*

*Cool-Season Forage Grasses*

*Corn Silage Production, Management, and Feeding*

*Forage Cell Wall Structure and Digestibility*

*Forage Quality, Evaluation, and Utilization*

*Pasture and Forage Crop Pathology*

*Persistence of Forage Legumes*

For information on these titles, please contact the ASA, CSSA, SSSA Headquarters Office; Attn.: Marketing; 677 South Segoe Road; Madison, WI 53711-1086. Telephone: (608) 273-8080, Fax: (608) 273-8089.

# Molecular and Cellular Technologies for Forage Improvement

Proceedings of a symposium sponsored by Divisions C-1, C-6, and C-7 of the Crop Science Society of America in Indianapolis, IN, 6 Nov. 1996.

*Editors*

E. C. Brummer, chair
N. S. Hill
C. A. Roberts

*Organizing Committee*

E. C. Brummer, co-chair
N. S. Hill, co-chair

*Editor-in-Chief CSSA*

J. J. Volenec

*Managing Editor*

David M. Kral

*Associate Editor*

Marian K. Viney

CSSA Special Publication Number 26

**Crop Science Society of America, Inc.**
**Madison, Wisconsin, USA**
**1998**

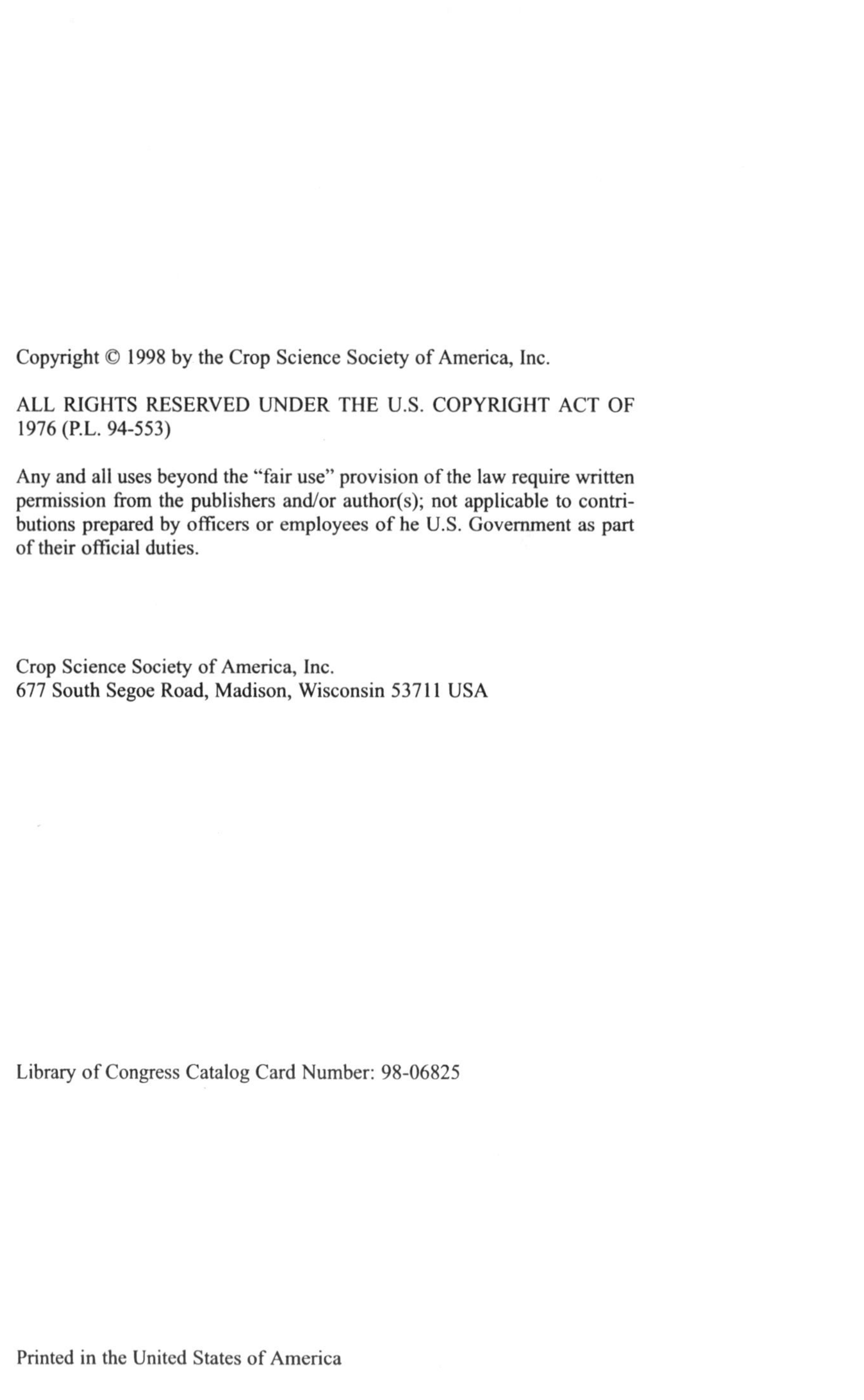

Cover: Photograph provided by Nicholas Hill;
cover design by Patricia Scullion.



Crop Science Society of America, Inc.
677 South Segoe Road, Madison, Wisconsin 53711 USA

Library of Congress Catalog Card Number: 98-06825

Printed in the United States of America

# CONTENTS

# FOREWORD

The improvement of crops depends on underlying scientific knowledge and the appropriate technologies to manipulate the species. Forage species often are complex genetically and require extra effort to apply molecular and cellular technologies. While forages offer many advantages to agriculture and opportunities for society, the research base has been difficult to maintain such that new technologies could be developed and applied in meaningful ways.

*Molecular and Cellular Technologies for Forage Improvement* is an impressive presentation of the spectrum of new information and technologies specifically developed for a spectrum of forage species. The development of molecular genetic maps and their applications in breeding, tissue culture technologies and their employment in transgenics, biochemical, and immunological approaches for improving forage species, and the novel use of endophytes illustrate the excitement that exists in the field today. The universality of much of the molecular and cellular information across species is a powerful force in shaping the opportunities available for the future.

This special publication documents that forage researchers are taking advantage of the new paradigms and offers an encouraging glimpse at how this is unfolding.

Ronald L. Phillips
*President-Elect Crop Science Society of America*
Regents' Professor, University of Minnesota
St. Paul, Minnesota

# PREFACE

The past decade has seen revolutionary advances in forage science. Forage crops have been transformed with novel genes offering herbicide resistances and improved forage quality characteristics. Molecular genetic linkage maps have been constructed in the major forages, and their use in marker assisted selection is beginning. Detailed understanding of the molecular basis of physiological processes such as winterhardiness and of plant-fungus symbioses is being reported, providing new avenues to improve forage crops. Despite this vibrant and innovative work, knowledge of the research outside a rather small cadre of forage scientists is limited. This special publication is intended to broaden the appreciation for this research, to help those not directly working with these technologies to gain a basic understanding of their implications, and to spur new collaborations and research ideas. I would like to thank Nick Hill, co-chair of the symposium and co-editor of the special publication, for initially suggesting we organize this effort and Craig Roberts for his editorial assistance. Thanks also to all the authors for timely and thought-provoking overviews describing their use of molecular and cellular technologies. Finally, thanks to the ASA headquarters staff, and particularly to Marian Viney, for production assistance.

E. CHARLES BRUMMER, *co-editor*
Iowa State University
Ames, Iowa

# CONTRIBUTORS

**Edwin T. Bingham** Professor, Department of Agronomy, University of Wisconsin, Madison, WI 53706

**Steve R. Bowley** Associate Professor, Department of Plant Agriculture, University of Guelph, Guelph, Ontario, Canada N1G 2W1

**Douglas J. Brouwer** Post Graduate Researcher, Department of Vegetable Crops, University of California, One Shields Ave., Davis, CA 95616

**E. Charles Brummer** Assistant Professor, Department of Agronomy, Iowa State University, Ames, IA 50011

**Yves Castonguay** Plant Physiologist, Agriculture and Agri-Food Canada, 2560 Hochelaga Boulevard, Sainte-Foy, Québec, Canada, G1V 2J3

**Tran C. Chanh** Senior Research Scientist, U.S. Army Medical Research Institute of Infectious Diseases, Ft. Detrick, Frederick, MD 21702

**C. Chen** Senior Research Associate, The Samuel Roberts Noble Foundation, 2510 Sam Noble Pkwy., Ardmore, OK 73402

**B.V. Conger** Professor, Department of Plant and Soil Science, University of Tennessee, Knoxville, TN 37901-1071

**S.M. Cunningham** Graduate Research Assistant, Department of Agronomy, Purdue University, West Lafayette, IN 47907-1150

**Emmett E. Hiatt, III** Postdoctoral Research Associate, Department of Crop and Soil Sciences, University of Georgia, Athens, GA 30602

**Nicholas S. Hill** Professor, Department Crop and Soil Sciences, University of Georgia, Athens, GA 30602

**B.C. Joern** Associate Professor, Department of Agronomy, Purdue University, West Lafayette, IN 47907-1150

**K.K. Kidwell** Assistant Professor, Department of Crop and Soil Science, Washington State University, Pullman, WA 99164-6420

**T.J. Kring** Professor, Department of Entomology, University of Arkansas, Fayetteville, AR 72701

**Serge Laberge** Molecular Biologist, Agriculture and Agri-Food Canada, 2560 Hochelaga Boulevard, Sainte-Foy, Québec, Canada, G1V 2J3

**M.L. Marlatt** Research Specialist, Department of Agronomy, University of Arkansas, 276 Altheimer Dr., Fayetteville, AR 72704

**M.E. McConnell** Biologist and Technical Information Specialist, American Type Culture Collection, 10801 University Boulevard, Manassas, VA 20110-2209

**Bryan D. McKersie** Professor, Department of Plant Agriculture, University of Guelph, Guelph, Ontario, Canada N1G 2W1

**Réal Michaud** Forage Plant Breeder, Agriculture and Agri-Food Canada, 2560 Hochelaga Boulevard, Sainte-Foy, Québec, Canada, G1V 2J3

**Paul Nadeau** Plant Biochemist, Agriculture and Agri-Food Canada, 2560 Hochelaga Boulevard, Sainte-Foy, Québec, Canada, G1V 2J3

**Thomas C. Osborn** — Professor, Department of Agronomy, University of Wisconsin, Madison, WI 53706

**A. Ourry** — U.A. INRA, Physiologie et Biochimie Vegetales, Institut de Recherche en Biologie Appliquee, Universite de Caen, 14032, Caen, Cedex, France

**E.L. Piper** — Professor, Department of Animal Science, University of Arkansas, Fayetteville, AR 72704

**D.A. Sleper** — Professor, Department of Agronomy, University of Missouri, Columbia, MO 65211

**Stefano Tavoletti** — Associate Professor, Dipartimento di Biotecnologie Agrarie ed Ambientali, Universita degli Studi di Ancona, 60131 Ancona, Italy

**Louis-P. Vézina** — Plant Biotechnologist, Agriculture and Agri-Food Canada, 2560 Hochelaga Boulevard, Sainte-Foy, Québec, Canada, G1V 2J3

**J.J. Volenec** — Professor, Department of Agronomy, Purdue University, West Lafayette, IN 47907-1150

**C.P. West** — Professor, Department of Agronomy, University of Arkansas, 276 Altheimer Dr., Fayetteville, AR 72704

# Conversion Factors for SI and non-SI Units

# Conversion Factors for SI and non-SI Units

| To convert Column 1 into Column 2, multiply by | Column 1 SI Unit | Column 2 non-SI Units | To convert Column 2 into Column 1, multiply by |
|---|---|---|---|
| | **Length** | | |
| 0.621 | kilometer, km ($10^3$ m) | mile, mi | 1.609 |
| 1.094 | meter, m | yard, yd | 0.914 |
| 3.28 | meter, m | foot, ft | 0.304 |
| 1.0 | micrometer, µm ($10^{-6}$ m) | micron, µ | 1.0 |
| $3.94 \times 10^{-2}$ | millimeter, mm ($10^{-3}$ m) | inch, in | 25.4 |
| 10 | nanometer, nm ($10^{-9}$ m) | Angstrom, Å | 0.1 |
| | **Area** | | |
| 2.47 | hectare, ha | acre | 0.405 |
| 247 | square kilometer, $km^2$ $(10^3\ m)^2$ | acre | $4.05 \times 10^{-3}$ |
| 0.386 | square kilometer, $km^2$ $(10^3\ m)^2$ | square mile, $mi^2$ | 2.590 |
| $2.47 \times 10^{-4}$ | square meter, $m^2$ | acre | $4.05 \times 10^{3}$ |
| 10.76 | square meter, $m^2$ | square foot, $ft^2$ | $9.29 \times 10^{-2}$ |
| $1.55 \times 10^{-3}$ | square millimeter, $mm^2$ $(10^{-3}\ m)^2$ | square inch, $in^2$ | 645 |
| | **Volume** | | |
| $9.73 \times 10^{-3}$ | cubic meter, $m^3$ | acre-inch | 102.8 |
| 35.3 | cubic meter, $m^3$ | cubic foot, $ft^3$ | $2.83 \times 10^{-2}$ |
| $6.10 \times 10^{4}$ | cubic meter, $m^3$ | cubic inch, $in^3$ | $1.64 \times 10^{-5}$ |
| $2.84 \times 10^{-2}$ | liter, L ($10^{-3}$ $m^3$) | bushel, bu | 35.24 |
| 1.057 | liter, L ($10^{-3}$ $m^3$) | quart (liquid), qt | 0.946 |
| $3.53 \times 10^{-2}$ | liter, L ($10^{-3}$ $m^3$) | cubic foot, $ft^3$ | 28.3 |
| 0.265 | liter, L ($10^{-3}$ $m^3$) | gallon | 3.78 |
| 33.78 | liter, L ($10^{-3}$ $m^3$) | ounce (fluid), oz | $2.96 \times 10^{-2}$ |
| 2.11 | liter, L ($10^{-3}$ $m^3$) | pint (fluid), pt | 0.473 |

| | | | |
|---|---|---|---|
| | **Mass** | | |
| $2.20 \times 10^{-3}$ | gram, g ($10^{-3}$ kg) | pound, lb | 454 |
| $3.52 \times 10^{-2}$ | gram, g ($10^{-3}$ kg) | ounce (avdp), oz | 28.4 |
| 2.205 | kilogram, kg | pound, lb | 0.454 |
| 0.01 | kilogram, kg | quintal (metric), q | 100 |
| $1.10 \times 10^{-3}$ | kilogram, kg | ton (2000 lb), ton | 907 |
| 1.102 | megagram, Mg (tonne) | ton (U.S.), ton | 0.907 |
| 1.102 | tonne, t | ton (U.S.), ton | 0.907 |
| | **Yield and Rate** | | |
| 0.893 | kilogram per hectare, kg $ha^{-1}$ | pound per acre, lb $acre^{-1}$ | 1.12 |
| $7.77 \times 10^{-2}$ | kilogram per cubic meter, kg $m^{-3}$ | pound per bushel, lb $bu^{-1}$ | 12.87 |
| $1.49 \times 10^{-2}$ | kilogram per hectare, kg $ha^{-1}$ | bushel per acre, 60 lb | 67.19 |
| $1.59 \times 10^{-2}$ | kilogram per hectare, kg $ha^{-1}$ | bushel per acre, 56 lb | 62.71 |
| $1.86 \times 10^{-2}$ | kilogram per hectare, kg $ha^{-1}$ | bushel per acre, 48 lb | 53.75 |
| 0.107 | liter per hectare, L $ha^{-1}$ | gallon per acre | 9.35 |
| 893 | tonnes per hectare, t $ha^{-1}$ | pound per acre, lb $acre^{-1}$ | $1.12 \times 10^{-3}$ |
| 893 | megagram per hectare, Mg $ha^{-1}$ | pound per acre, lb $acre^{-1}$ | $1.12 \times 10^{-3}$ |
| 0.446 | megagram per hectare, Mg $ha^{-1}$ | ton (2000 lb) per acre, ton $acre^{-1}$ | 2.24 |
| 2.24 | meter per second, m $s^{-1}$ | mile per hour | 0.447 |
| | **Specific Surface** | | |
| 10 | square meter per kilogram, $m^2$ $kg^{-1}$ | square centimeter per gram, $cm^2$ $g^{-1}$ | 0.1 |
| 1000 | square meter per kilogram, $m^2$ $kg^{-1}$ | square millimeter per gram, $mm^2$ $g^{-1}$ | 0.001 |
| | **Pressure** | | |
| 9.90 | megapascal, MPa ($10^6$ Pa) | atmosphere | 0.101 |
| 10 | megapascal, MPa ($10^6$ Pa) | bar | 0.1 |
| 1.00 | megagram, per cubic meter, Mg $m^{-3}$ | gram per cubic centimeter, g $cm^{-3}$ | 1.00 |
| $2.09 \times 10^{-2}$ | pascal, Pa | pound per square foot, lb $ft^{-2}$ | 47.9 |
| $1.45 \times 10^{-4}$ | pascal, Pa | pound per square inch, lb $in^{-2}$ | $6.90 \times 10^3$ |

(continued on next page)

# Conversion Factors for SI and non-SI Units

| To convert Column 1 into Column 2, multiply by | Column 1 SI Unit | Column 2 non-SI Units | To convert Column 2 into Column 1, multiply by |
|---|---|---|---|
| | | **Temperature** | |
| 1.00 (K – 273) | Kelvin, K | Celsius, °C | 1.00 (°C + 273) |
| (9/5 °C) + 32 | Celsius, °C | Fahrenheit, °F | 5/9 (°F – 32) |
| | | **Energy, Work, Quantity of Heat** | |
| $9.52 \times 10^{-4}$ | joule, J | British thermal unit, Btu | $1.05 \times 10^{3}$ |
| 0.239 | joule, J | calorie, cal | 4.19 |
| $10^{7}$ | joule, J | erg | $10^{-7}$ |
| 0.735 | joule, J | foot-pound | 1.36 |
| $2.387 \times 10^{-5}$ | joule per square meter, J $m^{-2}$ | calorie per square centimeter (langley) | $4.19 \times 10^{4}$ |
| $10^{5}$ | newton, N | dyne | $10^{-5}$ |
| $1.43 \times 10^{-3}$ | watt per square meter, W $m^{-2}$ | calorie per square centimeter minute (irradiance), cal $cm^{-2}$ $min^{-1}$ | 698 |
| | | **Transpiration and Photosynthesis** | |
| $3.60 \times 10^{-2}$ | milligram per square meter second, mg $m^{-2}$ $s^{-1}$ | gram per square decimeter hour, g $dm^{-2}$ $h^{-1}$ | 27.8 |
| $5.56 \times 10^{-3}$ | milligram ($H_2O$) per square meter second, mg $m^{-2}$ $s^{-1}$ | micromole ($H_2O$) per square centimeter second, µmol $cm^{-2}$ $s^{-1}$ | 180 |
| $10^{-4}$ | milligram per square meter second, mg $m^{-2}$ $s^{-1}$ | milligram per square centimeter second, mg $cm^{-2}$ $s^{-1}$ | $10^{4}$ |
| 35.97 | milligram per square meter second, mg $m^{-2}$ $s^{-1}$ | milligram per square decimeter hour, mg $dm^{-2}$ $h^{-1}$ | $2.78 \times 10^{-2}$ |
| | | **Plane Angle** | |
| 57.3 | radian, rad | degrees (angle), ° | $1.75 \times 10^{-2}$ |

| | | | |
|---|---|---|---|
| | **Electrical Conductivity, Electricity, and Magnetism** | | |
| 10 | siemen per meter, S $m^{-1}$ | millimho per centimeter, mmho $cm^{-1}$ | 0.1 |
| $10^4$ | tesla, T | gauss, G | $10^{-4}$ |
| | **Water Measurement** | | |
| $9.73 \times 10^{-3}$ | cubic meter, $m^3$ | acre-inches, acre-in | 102.8 |
| $9.81 \times 10^{-3}$ | cubic meter per hour, $m^3$ $h^{-1}$ | cubic feet per second, $ft^3$ $s^{-1}$ | 101.9 |
| 4.40 | cubic meter per hour, $m^3$ h−1 | U.S. gallons per minute, gal $min^{-1}$ | 0.227 |
| 8.11 | hectare-meters, ha-m | acre-feet, acre-ft | 0.123 |
| 97.28 | hectare-meters, ha-m | acre-inches, acre-in | $1.03 \times 10^{-2}$ |
| $8.1 \times 10^{-2}$ | hectare-centimeters, ha-cm | acre-feet, acre-ft | 12.33 |
| | **Concentrations** | | |
| 1 | centimole per kilogram, cmol $kg^{-1}$ | milliequivalents per 100 grams, meq 100 $g^{-1}$ | 1 |
| 0.1 | gram per kilogram, g $kg^{-1}$ | percent, % | 10 |
| 1 | milligram per kilogram, mg $kg^{-1}$ | parts per million, ppm | 1 |
| | **Radioactivity** | | |
| $2.7 \times 10^{-11}$ | becquerel, Bq | curie, Ci | $3.7 \times 10^{10}$ |
| $2.7 \times 10^{-2}$ | becquerel per kilogram, Bq $kg^{-1}$ | picocurie per gram, pCi $g^{-1}$ | 37 |
| 100 | gray, Gy (absorbed dose) | rad, rd | 0.01 |
| 100 | sievert, Sv (equivalent dose) | rem (roentgen equivalent man) | 0.01 |
| | **Plant Nutrient Conversion** | | |
| | ***Elemental*** | ***Oxide*** | |
| 2.29 | P | $P_2O_5$ | 0.437 |
| 1.20 | K | $K_2O$ | 0.830 |
| 1.39 | Ca | CaO | 0.715 |
| 1.66 | Mg | MgO | 0.602 |

# 1 Molecular and Cellular Technologies in Forage Improvement: An Overview[1]

**E. Charles Brummer**

*Department of Agronomy*
*Iowa State University*
*Ames, Iowa*

Forage crops represent essential components of North American agricultural systems. The negative consequences of long-term row crop monocultures, such as soil degradation and erosion, diminished aquifers and lower water quality, and persistent weed and pest problems, are among society's most pressing problems and are being widely debated (Matson et al., 1997). Many of these problems can be ameliorated by integrating forages into a crop rotation. Forage crops limit soil loss, improve water infiltration and quality, and promote sustainable crop production, which in turn can boost farm profitability and diversification. Forages form the backbone of the beef, dairy, sheep, and horse industries and could play important roles in both hog and poultry production. The importance of forage crops extends beyond feed; they also are grown as roadside plantings, for revegetation on disturbed sites, and for biofuels. Some forages may eventually serve as perennial grain crops, providing human food in addition to livestock forage (Jackson, 1980). Because forages are not directly consumed by humans, their true worth is often underestimated. In a recent accounting of the benefits of various ecosystems, the value of grasslands was estimated to be 2.5 times greater than that of cropland (Costanza et al., 1997). Given the multiple uses of forage crops and their underlying worth to our agricultural system and our environment, research programs to improve the versatility of these crops are clearly warranted.

Forage improvement may be realized through better genetics, a better understanding of physiology or biochemistry, or more scientific management practices. Traits of interest vary by geographical region, crop species, livestock species, and management strategy, including enhanced persistence for long-term permanent pastures to better forage quality of corn (*Zea mays* L.) silage. Improvement strategies have become increasingly complex. As an example, forage breeding methodology has progressed from little more than natural selection

[1] Contribution of the Iowa Agric. and Home Econ. Exp. Stn., Ames, IA 50011. Project no. 2569. Journal Paper no. J-17729.


*Molecular and Cellular Technologies for Forage Improvement*. CSSA Special Publication no. 26.

(e.g., Webber, 1912) to the more structured approaches of progeny testing in use today (e.g., Vogel & Pederson, 1993; Rumbaugh et al., 1988). Mechanization of plot equipment has allowed breeders to evaluate more genotypes in multilocation trials. Statistical theory has provided better plot designs, such as α-lattices (Patterson & Williams, 1976), to improve the precision of selection. Finally, laboratory methods, including near infrared reflectance spectroscopy (NIRS; Shenk & Westerhaus, 1994), have vastly increased sample throughput, further allowing breeders to evaluate larger populations for desirable plants.

Even though these and other improved techniques during the past 100 yr have allowed forage researchers to enhance the productivity and usefulness of forage crops, new methods need to be applied to forage breeding as we enter the 21st century for continued improvement of a number of traits, namely yield, persistence, and quality. Some of the more intractable problems of forage improvement still exist: increasing productivity without sacrificing winter survival, improving quality without decreasing yield, developing germplasm for compatibility with other species, assessing the importance of diversity at the allelic and species levels on temporal and spatial yield stability, identifying the extent of genetic erosion in forage germplasm, and managing symbionts (e.g., *Neotyphodium* and *Rhizobium*) for optimal interactions are just a sampling of areas that need more research and development. The new suite of technologies that will be important in realizing some of these goals include the molecular and cellular tools developed over the past several decades.

Many of the technologies discussed in this volume have been used to a much greater extent in other crop species or in other living organisms than in the forages. For instance, restriction fragment length polymorphisms (RFLP), a class of molecular markers, were proposed for human genetic mapping in 1980 (Botstein et al., 1980); the first RFLP plant map was described in tomato [*Lycopersicon lycopersicum* (L.) Karsten] in 1986 (Bernatzky & Tanksley, 1986); and not until 1993 were maps developed for alfalfa (*Medicago sativa* L.), the first forage crop to be mapped (Brummer et al., 1993; Kiss et al., 1993). Similar histories can be traced for the other technologies as well. Forage crops often lag behind the better funded grain and vegetable crops due to poor lobbying for research monies as well as to biological difficulties such as polyploidy, perenniality, and severe inbreeding depression; however, forage researchers can adopt this technology for use on their species particularly after initial difficulties are overcome in other crops.

Molecular and cellular technologies represent the next successional development in better research equipment, tools that augment, not supplant, existing technology and research procedures. Far from eliminating the need for the breeder or the physiologist, these technologies promise to make their jobs more effective. The chapters in this publication were presented at a symposium designed to discuss how some of these technologies are being used in current research and to stimulate thinking within the forage community about their expanded use in other programs. The authors have all made substantial contributions in their fields, but they are not alone in using biotechnology in forage crops. Numerous groups are using mapping, transformation, gene cloning, and other techniques to improve crops for various purposes.

## USES OF MOLECULAR TECHNOLOGIES

Much forage research culminates directly or indirectly in the development of new cultivars better suited to a particular cropping system than their predecessors. By relying mainly on whole plant phenotypic evaluation, traditional techniques do not directly assess a plant's genotype, at least for the majority of agronomically important traits. Elucidation of the biochemistry and genetics of forage crops enables a more refined and precise evaluation of plants thereby improving the efficiency of cultivar synthesis. With the technologies discussed in this symposium, we are gaining the ability to make critical improvements in understanding the cellular and molecular basis of forage crop growth and development as well as in devising superior methods for crop improvement.

### Germplasm Identification and Characterization

The key to a successful cultivar development program revolves around genetic diversity. Initially, breeders must identify germplasm sources with sufficient variation for traits of interest in order to make selections. Germplasm sources can be obtained from essentially three areas: (1) adapted, previously selected material, (2) unadapted, exotic accessions that may or may not have undergone some selection, and (3) native or naturalized, unselected plants. In the early stages of a selection program for a species, Groups 2 and 3 may provide the best starting points; in fact, no previously selected material may be available; however, as selection progresses, the first group will become the predominant source of breeding material, with the other groups serving as reserves if the need for particular traits arises.

Molecular tools are increasingly being used to characterize germplasm in an attempt to guide selection of appropriate material for testing, to help formulate future breeding efforts, or to curate plant introduction collections (e.g., Brummer et al. 1991, 1995; Huff et al., 1997; Mitchell et al., 1997). Molecular markers may help target appropriate accessions that could be useful for particular evaluation programs making field nurseries more manageable and effective. Markers could help truncate a germplasm collection by identifying unique accessions, i.e., ones with novel alleles not present in cultivated material, or by limiting redundancy while still ensuring that the full range of genetic diversity is evaluated. Markers can assist germplasm curators in developing comprehensive collections by eliminating duplications and identifying deficiencies (Mitchell et al., 1997).

Germplasm available for crop improvement today has a considerably broader range than even in the recent past. Until recent biotechnological developments, the usable germplasm pool consisted of cultivated material and those wild relatives with which it could hybridize in some manner. Harlan and deWet (1971) identified three germplasm pools—primary, secondary, and tertiary—to which, with the advent of molecular technologies, could be added the *quaternary gene pool* based on procedures such as genetic transformation and somatic hybridization. Transformation technologies potentially allow any cloned gene, regardless of its source, to be inserted into a crop plant. Conger (1998, this publication) discusses some of the strategies and pitfalls of transforming forage grasses. Considerable work also has been done on forage legumes, e.g., alfalfa (Tabe, 1995; Austin & Bingham, 1997; Dixon et al., 1996), white clover (*Trifolium repens* L.; White & Greenwood, 1987; Larkin et al., 1996), subterranean clover (*T. subterraneum* L.; Khan et al., 1994), and birdsfoot trefoil (*Lotus corniculatus* L.; Webb et al., 1996). Other grass research, primarily with the fescues (*Festuca* sp.) and ryegrasses (*Lolium* sp.), also is ongoing (e.g., Wang et al., 1992). Dipping into the growing cloned gene pool may provide forages with improved quality, pest resistances, nutrient requirements, and other traits.

Somatic hybridization, in which genomes of two distantly related species possibly from the secondary or tertiary gene pools are combined via protoplast fusion, has opened up a panoply of possibilities for gene transfer and crop improvement. Successful somatic hybrids between genera such as *Festuca* and *Lolium* (Spangenberg, 1994) and *Medicago* and *Onobrychis* (Li et al., 1993) have been reported. Arcioni et al. (1997) provide an extensive review of somatic hybridization in the forage legumes. Thus, the current germplasm pool of forage crops is quite extensive and allows broad possibilities of producing altered forage cultivars.

## Selection and Population Development

Any germplasm source, whether plant introductions or introduced genes, must go through a breeding program to put the desired traits into an adapted, agronomically acceptable package. The biggest problem of conventional plant breeding is that selection is based primarily on phenotype. Expression of many agronomic traits (e.g., winterhardiness, yield, or quality) is modulated by the environment in which the plants are grown so that even multi-environment evaluations and progeny test data only approximate the true genotype of a plant. For some traits, low heritabilities, resulting from high genotype × environment interactions, may result in little or no genetic gain from selection. Further hampering gain in the forage crops is their perennial nature, which necessitates multiyear evaluations, particularly for traits such as winterhardiness and persistence. Selection based on seedling assays would considerably accelerate breeding cycles.

Basing selection decisions on molecular or cellular marker results can limit environmental noise and allow more accurate ascertainment of the true genotype of the plant. Using a molecular marker for the gene(s) controlling a particular trait increases heritability to unity (or nearly so) if the marker is extremely tightly linked to the trait. In this respect, marker-facilitated backcrossing of transgenes, where a marker for the transgene itself is readily available, offers a simple and effective method for introgression. For complex traits, even small increases in the precision of genotype identification can lead to significant improvements in the trait under selection (Tanksley et al., 1996; Stuber et al., 1992).

The basic idea of marker-facilitated selection is simple. In conventional selection, breeders select individuals in one (or if divergent selection, both) of the tails of a typical Gaussian distribution for some quantitative trait, perhaps the upper 5% of the population. These individuals are phenotypically superior and, if the trait is highly heritable, and they will share specific alleles affecting the trait.

Because we know that many traits are not highly heritable (e.g., yield), the individuals in the upper 5% may not share the desired alleles. Because the ultimate goal is to select the superior *alleles* in the population, protocols that ensure selection of particular genotypes will avoid dilution of the selected population with suboptimal individuals. If a marker were previously linked to the trait of interest, we could select plants based on the desired marker, *whether or not they were in the tail of the phenotypic distribution*. Thus, selecting the marker rather than (or in conjunction with) the phenotype results in improved genetic gain. Perhaps the best approach is to combine phenotypic evaluation with marker data, rather than relying on marker data exclusively (Lande & Thompson, 1990).

Success of marker-assisted selection schemes depend on good markers for the traits of interest. For quantitative trait loci, molecular markers will be most useful in aiding the breeding program due to their abundance and even distribution throughout the genome; however, a saturated molecular linkage map for the population under study is essential for marker-assisted selection to be useful. Osborn et al. (1998, this publication) and Sleper and Chen (1998, this publication) discuss the status of molecular mapping in alfalfa and tall fescue (*F. arundinacea* Schreber), respectively. Though the maps are still in a relatively nascent stage compared with other crops, these two chapters indicate that significant research is ongoing, at least in these two economically important species. As the chapters suggest, we are still a considerable distance from routine marker usage for cultivar improvement in most forages.

Markers also may be useful for simply inherited traits that are difficult to screen or score. Developing markers that can be readily assayed or techniques that can speed screening dramatically will improve breeding efficiency. Hill et al. (1998, this publication) discuss the use of immunochemical methods to detect the presence of ergot alkaloids in tall fescue. Monoclonal antibody screening can greatly increase the efficiency of detection and selection for particular chemical constituents, expediting seed lot evaluations and breeding programs. Hiatt et al. (1997) also describe a rapid ELISA method to detect the presence of *Neotyphodium coenophialum* in tall fescue plants, further improving the efficiency of fescue breeding. Likewise, Castonguay et al. (1997; 1998, this publication) and Volenec et al. (1998, this publication) discuss experiments to evaluate plants for winterhardiness, by studying various constituents such as proteins, sugars, and mRNA transcripts. Further work in these areas will undoubtedly shed more light on the causes of winterhardiness and lead to better options in breeding for this essential trait without sacrificing yield.

Waiting to make selections based on evaluations of adult plants severely restricts forage breeding programs. Cycles of selection would be markedly faster if selection could be practiced in juvenile plants rather than waiting for full trait expression in adults. For forages, seedling selection could eliminate years per cycle of evaluation, potentially allowing multiple cycles to be completed for every one in a traditional program, if the markers are reliably linked to the trait. Selection in the seedling stage could possibly identify individuals with traits of interest, or alternatively, markers could be used to discard plants that are definitely not useful. Selecting the desired phenotypes requires strong links of markers to traits; whether these strong associations will become feasible for forages is not clear. A loose association of marker and trait could, however, clearly aid in discarding undesirable plants, allowing breeders either to limit the size of field selection nurseries or to increase the effective size of the nursery by only evaluating plants that have better potential of being superior genotypes. Two putative single-gene systems that would benefit greatly from tightly linked molecular markers are apomixis (e.g., Gustine et al., 1996) and rhizome production in birdsfoot trefoil (Beuselinck & Steiner, 1996). In both cases, a considerable breeding effort could be saved by culling seedlings rather than adult plants.

Because most forage cultivars are synthetic populations and because most forages are negatively affected by inbreeding, selecting highly diverse parents could minimize inbreeding depression, or maximize heterosis, in the resulting cultivar (Bingham et al., 1994). Simple genetic diversity has not adequately identified highly heterotic hybrids in most cases (e.g., Bernardo, 1992; Godshalk et al., 1990); however, preselecting markers that have been associated with heterotic regions as the basis of a diversity analysis has been more successful in rice (Zhang et al., 1994). A similar approach in forage crops is underway and should prove useful in developing highly heterotic synthetics (Brummer, 1998, unpublished data).

## Improving Symbioses

Until this point, we have mainly been concerned with selection in plant populations; however, in a number of situations, notably *Festuca–Lolium-Neotyphodium* and legume–*Rhizobium* associations, the symbiont species also needs to be considered during the selection process. Due to the severe animal disorders caused by *Neotyphodium* sp. (Hoveland, 1993), considerable interest in altering the fungus has recently developed. West et al. (1998, this publication) discuss some of the background of this approach and discuss potential research directions. The legume–*Rhizobium* interaction has continues to attract considerable attention. Barran and Bromfield (1997) consider some of the possibilities in genetically engineering *Rhizobium* sp. to improve N fixation. Clearly, both of these symbionts play a large role in their respective associations and need to be examined for their impact on improving forage crops.

## Cultivar Development

Alternatives to the commonly used synthetic methods of cultivar production could provide better ways to capitalize on heterosis. Hybrid alfalfa has been attempted several times in the past, but logistic and cost problems have limited its commercialization (Viands et al., 1988). McKersie and Bowley (1998, this publication) discuss some of the possibilities of using somatic embryos as artificial seeds to develop hybrid or reduced generation synthetic alfalfa. These methods would capture more heterosis than conventional synthetics and may result in significant yield increases. The effect of the limited genetic variation on yield stability is an important consideration in development of these cultivars and needs to be addressed. Nevertheless, these new technologies will allow greater flexibility in producing different types of cultivars.

Preserving identity and proprietary rights of cultivars are also of great interest, particular to commercial breeding companies. Molecular markers have assisted in this regard (e.g., Huff, 1997), though more research is needed to prove that population differentiation is possible. Though fingerprinting heterogeneous and heterozygous populations is more difficult than inbred lines or hybrids, success of the process should be possible.

## ABUSES OF MOLECULAR TECHNIQUES

Molecular techniques offer substantial advantages over traditional improvement methods; however, not all situations need the power of these methods. A potential problem with the availability of these techniques is that their use may divert resources from other, more important problems that could be solved by traditional techniques. Our research goals should be the improvement of environmental and economic stability of our agricultural systems. Forages are a complex of species: thus, a deficiency in one species may be easily overcome by substituting a different species rather than attempting a difficult biotechnological solution. Canalization of research focus on a single species is common throughout agricultural research, but broadening our focus to the overarching problem of agricultural productivity and profitability may represent a better approach to crop improvement in some instances. One example of this is bloat in alfalfa. Is using biotechnology to develop bloat safe alfalfa a better approach than traditional breeding of nonbloating legumes such as birdsfoot trefoil? Though conventional breeding has limitations, a concerted effort will result in improved cultivars, possibly at lower cost than biotechnology efforts in alfalfa. Bloat-safe alfalfa would be an important management tool for livestock and dairy producers, but limited financial resources may be better allocated in other ways. Given that financial restrictions are inevitable, our focus should concentrate on overcoming problems that limit forages as a whole. Practical solutions should be sought before embarking on costly biotechnological approaches.

## CONCLUDING REMARKS

Forage research can dramatically improve a farmer's profitability. Several reports have shown that small improvements in forage digestibility improved animal gain (Anderson et al., 1988; Eichorn et al., 1986; Moore et al., 1995). Newer alfalfa varieties routinely yield more than older, check cultivars such as Vernal, Riley, and Saranac AR. Three-year average yields of the four top performing cultivars in the 1993 alfalfa variety test at Ames, IA, were 127% higher than the check cultivars (13.7 vs. 10.8 Mg ha$^{-1}$, Brummer & Crim, 1996). Molecular technologies will be essential for additional improvements in traits that are difficult to quantify.

Simply because molecular and cellular technologies can be powerful tools in forage research does not mean all research programs would benefit from their use; however, researchers need to be aware of their availability and utility. As a forage community, collaborative arrangements between traditional agronomists and molecular biologists or cellular chemists seem less frequent than in other commodity areas; however forage scientists are being trained with these technologies, which should aid in their widespread adoption by the forage community.

Despite the importance of forages, forage research suffers from funding restrictions due, in part, to the lack understanding of their contribution to the farm economy and environmental quality. Molecular and cellular technologies discussed in the ensuing chapters have the potential to further improve forage production and profitability, making the integration of forages into our farming systems more compelling.

## REFERENCES

Anderson, B., J.K. Ward, K.P. Vogel, M.G. Ward, H.J. Gorz, and F.A. Haskins. 1988. Forage quality and performance of yearlings grazing switchgrass strains selected for differing digestibility. J. Anim. Sci. 66:2239–2244.

Arcioni, S., F. Damiani, A. Mariani, and F. Pupilli. 1997. Somatic hybridization and embryo rescue for the introduction of wild germplasm. p. 61–89. *In* B.D. McKersie and D.C.W. Brown (ed.) Biotechnology and the improvement of forage legumes. CAB Int., Wallingford, England.

Austin, S., and E.T. Bingham. 1997. The potential use of transgenic alfalfa as a bioreactor for the production of industrial enzymes. p. 409–424. *In* B.D. McKersie and D.C.W. Brown (ed.) Biotechnology and the improvement of forage legumes. CAB Int., Wallingford, England.

Barran, L.R., and E.S.P. Bromfield. 1997. Competition among *Rhizobia* for nodulation of legumes. p. 343–374. *In* B.D. McKersie and D.C.W. Brown (ed.) Biotechnology and the improvement of forage legumes. CAB Int., Wallingford, England.

Bernatzky, R., and S.D. Tanksley. 1986. Toward a saturated linkage map in tomato based on isozymes and random cDNA sequences. Genetics 112:887–898.

Bernardo, R. 1992. Relationship between single-cross performance and molecular marker heterozygosity. Theor. Appl. Genet. 83:628–634.

Beuselinck, P.R., and J.J. Steiner. 1996. Registration of 'ARS-2620' birdsfoot trefoil. Crop Sci. 36:1414.

Bingham, E.T., R.W. Groose, D.R. Woodfield, and K.K. Kidwell. 1994. Complementary gene interactions in alfalfa are greater in autotetraploids than diploids. Crop Sci. 34:823–829.

Botstein, D., R.L. White, M. Skolnick, and R.W. Davis. 1980. Construction of a genetic linkage map in man using restriction fragment length polymorphisms. Am. J. Hum. Genet. 32:314–331.

Brummer, E.C., J.H. Bouton, and G. Kochert. 1993. Development of an RFLP map in diploid alfalfa. Theor. Appl. Genet. 86:329–332.

Brummer, E.C., J.H. Bouton, and G. Kochert. 1995. Analysis of annual *Medicago* species using RAPD markers. Genome 38:362–367.

Brummer, E.C., and L. Crim. 1996. 1996 Alfalfa yield test report. Iowa State Univ. Ext. Bull. AG-84. Iowa State Univ., Ames.

Brummer, E.C., G. Kochert, and J.H. Bouton. 1991. RFLP variation in diploid and tetraploid alfalfa Theor. Appl. Gen. 83:89–96.

Castonguay, Y., S. Laberge, R. Michaud, P. Nadeau, and L.P. Vézina. 1998. Molecular approaches for the improvement of cold tolerance in alfalfa. p. 33–48. *In* Molecular and cellular technologies for forage improvement. CSSA Spec. Publ. 26. CSSA, Madison, WI (this publication).

Castonguay, Y., P. Nadeau, S. Laberge, and L.P. Vézina. 1997. Changes in gene expression in six alfalfa cultivars acclimated under winter hardening conditions. Crop Sci. 37:332–342.

Costanza, R., R. d'Arge, R. de Groot, S. Farber, M. Grasso, B. Hannon, K. Limburg, S. Naeem, R.V. O'Neill, J. Paruelo, R.G. Raskin, P. Sutton, and M. van den Belt. 1997. The value of the world's ecosystem services and natural capital. Nature (London) 387:253–260.

Conger, B.V. 1998. Genetic transformation of forage grasses. p. 49–58. *In* Molecular and cellular technologies for forage improvement. CSSA Spec. Publ. 26. CSSA, Madison, WI (this publication).

Dixon, R.A., C.J. Lamb, S. Masoud, V.J.H. Sewalt, and N.L. Paiva. 1996. Metabolic engineering: Prospects for crop improvement through the genetic manipulation of phenylpropanoid biosynthesis and defense responses—a review. Gene 179:61–71.

Eichorn, M.M., Jr., W.M. Oliver, W.B. Hallmark, W.A. Young, A.V. Davis, and B.C. Nelson. 1986. Registration of Grazer' bermudagrass. Crop Sci. 26:835.

Godshalk, E.B., M. Lee, and K.R. Lamkey. 1990. Relationship of restriction fragment length polymorphisms to single-cross hybrid performance of maize. Theor. Appl. Genet. 80:273–280.

Gustine, D.L., R.T. Sherwood, Y. Gounaris, and D. Huff. 1996. Isozyme, protein, and RAPD markers within a half-sib family of buffelgrass segregating for apospory. Crop Sci. 36:723–727.

Harlan, J.R., and J.M.J. deWet. 1971. Toward a rational classification of cultivated plants. Taxon 20:509–517.

Hiatt, E.H., III, N.S. Hill, J.H. Bouton, and C.W. Mims. 1997. Monoclonal antibodies for detection of *Neotyphodium coenophialum*. Crop Sci. 37:1265–1269.

Hill, N.S., E.E. Hiatt III, and T.C. Chanh. 1998. Immunological Approaches for Improving Forage Species. p. 75–104. *In* Molecular and cellular technologies for forage improvement. CSSA Spec. Publ. 26. CSSA, Madison, WI (this publication).

Hoveland, C.S. 1993. Importance and economic significance of the *Acremonium* endophyte to performance of animals and the grass plant. Agric. Ecosyst. Environ. 44:3–12.

Huff, D.R. 1997. RAPD characterization of heterogeneous perennial ryegrass cultivars. Crop Sci. 37:557–564.

Jackson, W. 1980. New roots for agriculture. Univ. of Nebraska Press, Lincoln.

Kahn, M.R.I., L.M. Tabe, L.C. Heath, D. Spencer, and T.J.V. Higgins. 1994. Agrobacterium-mediated transformation of subterranean clover (*Trifolium subterraneum* L.). Plant Physiol. 105:81–88.

Kiss, G.B., G. Csanádi, K. Kálmán, P. Kaló, and L. Ökrész. 1993. Construction of a basic genetic map for alfalfa using RFLP, RAPD, isozyme and morphological markers. Mol. Gen. Genet. 238:129–137.

Larkin, P.J., J.M. Gibson, U. Mathesius, E. Gartner, E. Hall, G.J. Tanner, B.G. Rolfe, and M.A. Djordjevic. 1996. Transgenic white clover studies with the auxin responsive promoter, GH3, in root gravitropism and lateral root development. Transgen. Res. 5:325–335.

Lande, R., and R. Thompson. 1990. Efficiency of marker-assisted selection in the improvement of quantitative traits. Genetics 124:743–756.

Li, Y-G., G.J. Tanner, A.C. Delves, and P.J. Larkin. 1993. Asymmetric somatic hybrid plants between *Medicago sativa* L. (alfalfa, lucerne) and *Onobrychis viciifolia* Scop. (sainfoin). Theor. Appl. Genet. 87:455–463.

Matson, P.A., W.J. Parton, A.G. Power, and M.J. Swift. 1997. Agricultural intensification and ecosystem properties. Science (Washington, DC) 277:504–509.

McKersie, B.D., and S.R. Bowley. 1998. Somatic embryogenesis: Forage improvement using synthetic seeds and plant transformation. p. 117–134. *In* Molecular and cellular technologies for forage improvement. CSSA Spec. Publ. 26. CSSA, Madison, WI (this publication).

Mitchell, S.E., S. Kresovich, C.A. Jester, C.J. Hernandez, and A.K. Szewc-McFadden. 1997. Application of multiplex PCR and fluorescence-based, semi automated allele sizing technology for genotyping plant genetic resources. Crop Sci. 37:617–624.

Moore, K.J., K.P. Vogel, T.J. Klopfenstein, and B.E. Anderson. 1995. Evaluation of four intermediate wheatgrass populations under grazing. Agron. J. 87:744–747.

Osborn, T.C., D.B. Brouwer, K.K. Kidwell, S. Tavoletti, and E.T. Bingham. 1998. Molecular marker applications to genetics and breeding of alfalfa. p. 25–31. *In* Molecular and cellular technologies for forage improvement. CSSA Spec. Publ. 26. CSSA, Madison, WI (this publication).

Patterson, H.D., and E.R. Williams. 1976. A new class of resolvable incomplete block designs. Biometrika 63:83–92.

Rumbaugh, M.D., J.L. Caddel, and D.E. Rowe. 1988. Breeding and quantitative genetics. p. 777–808. *In* A.A. Hanson et al. (ed.) Alfalfa and alfalfa improvement. ASA, CSSA, and SSSA, Madison, WI.

Shenk, J.S., and M.O. Westerhaus. 1994. The application of near infrared reflectance spectroscopy (NIRS) to forage analysis. p. 406–449. *In* G.C. Fahey, Jr. (ed.) Forage quality, evaluation, and utilization. ASA, CSSA, and SSSA, Madison, WI.

Sleper, D.A., and C. Chen. 1998. Molecular mapping of forage grasses. p. 11–24. *In* Molecular and cellular technologies for forage improvement. CSSA Spec. Publ. 26. CSSA, Madison, WI (this publication).

Spangenberg, G., M.P. Valles, Z.Y. Wang, P. Montavon, J. Nagel, and I. Potrykus. 1994. Asymmetric somatic hybridization between tall fescue (*Festuca arundinacea* Schreb.) and irradiated Italian ryegrass (*Lolium multiflorum* Lam.) protoplasts. Theor. Appl. Genet. 88:509–519.

Stuber, C.W., S.E. Lincoln, D.W. Wolff, T. Helentjaris, and E.S. Lander. 1992. Identification of genetic factors contributing to heterosis in a hybrid from two elite maize inbred lines using molecular markers. Genetics 132:823–839.

Tabe, L.M., T. Wardley-Richardson, A. Ceriotti, A. Aryan, W. McNabb, A. Moore, and T.J.V. Higgins. 1995. A biotechnological approach to improving the nutritive value of alfalfa. J. Anim. Sci. 73:2752–2759.

Tanksley, S.D., S. Grandillo, T.M. Fulton, D. Zamir, Y. Eshed, V. Petiard, J. Lopez, and T. Beck-Bunn. 1996. Advanced backcross QTL analysis in a cross between an elite processing line of tomato and its wild relative *L. pimpinellifolium*. Theor. Appl. Genet. 92:213–224.

Viands, D.R., P. Sun, and D.K. Barnes. 1988. Pollination control: Mechanical and sterility. p. 931–960. *In* A.A. Hanson et al. (ed.) Alfalfa and alfalfa improvement. ASA, CSSA, and SSSA, Madison, WI.

Vogel, K.P., and J.F. Pederson. 1993. Breeding systems for cross-pollinated perennial grasses. Plant Breed. Rev. 11:251–274.

Volenec, J.J., B.C. Joern, A. Ourry, and S.M. Cunninham. 1998. Molecular analysis of alfalfa root vegetative storage proteins. p. 59–73. *In* Molecular and cellular technologies for forage improvement. CSSA Spec. Publ. 26. CSSA, Madison, WI (this publication).

Wang, Z.Y., T. Takamizo, V.A. Iglesias, M. Osusky, J. Nagel, I. Potrykus, and G. Spangenberg. 1992. Transgenic plants of tall fescue (*Festuca arundinacea* Schreb.) obtained by direct gene transfer to protoplasts. Bio/Technol. 10:691–696.

Webb, K.J., M.J. Gibbs, S. Mizen, L. Skot, and J.A. Gatehouse. 1996. Genetic transformation of *Lotus corniculatus* with *Agrobacterium tumifaciens* and the analysis of the inheritance of transgenes in the T-1 generation. Transgen. Res. 5:303–312.

Webber, H.J. 1912. The production of new and improved varieties of timothy. p. 337–392. New York (Cornell) Agric. Exp. Stn. Bull. 313. Cornell Univ., Ithaca, NY.

West, C.P., E.L. Piper, M.L. Marlott, M.E. McConnell, and T.J. Kring. 1998. Novel endophyte technology: Selection of the fungus. p. 105–115. *In* Molecular and cellular technologies for forage improvement. CSSA Spec. Publ. 26. CSSA, Madison, WI (this publication).

White, D.W.R., and D. Greenwood. 1987. Transformation of the forage legume *Trifolium repens* L. using binary *Agrobacterium* vectors. Plant Molec. Biol. 8:461–469.

Zhang, Q., Y.J. Gao, S.H. Yang, R.A. Ragab, M.A. Saghai-Maroof, and Z.B. Li. 1994. A diallel analysis of heterosis in elite hybrid rice based on RFLPs and microsatellites. Theor. Appl. Genet. 89:185–192.

# 2 Molecular Mapping of Forage Grasses

**D. A. Sleper and C. Chen**

*Department of Agronomy*
*University of Missouri*
*Columbia, Missouri*

## ABSTRACT

Mapping of forage grasses using molecular markers is occurring for several different species; however, mapping of forage grasses lags behind that of many other crop plants. It is important that mapping occur for forage grasses as it is expected to provide insights into the organization and evolution of plant genomes. This is important for forage grasses because most o the species are highly polyploid. Molecular maps can provide information on assaying genetic variation within a species. In addition, important genes for desirable traits can be localize by linkage analysis, cloned on the basis of a linkage map, and transformed to produce improved cultivars. It is also hoped that well developed linkage maps can identify useful quantitative trait loci that could be used in marker-assisted selection for forage grass improvement.

Molecular mapping of crop plants is occurring in many research laboratories throughout the world. Most of the major crops are being mapped with molecular markers, including for example, maize (*Zea mays* L.; Beavis & Grant, 1991), rye ( *Secale cereale* L.; Korzun et al., 1996), alfalfa (*Medicago sativa* L.; Brummer et al., 1993), rice (*Oryza sativa* L.; Kilian et al., 1995), tomato [*Lycopersicon lycopersicum* (L.) Karstenp; Chagué et al., 1996], barley (*Hordeum vulgare* L.; Kleinhofs et al., 1993), wheat (*Triticum aestivum* L.; Hart, 1994), oat (*Avena sativa* L.; O'Donoughue et al., 1994), soybean [*Glycine max* (L.) Merr.; Webb et al., 1995], sorghum [*Sorghum bicolor* (L.) Moench; Berhan et al., 1993], tall fescue (*Festuca arundinacea* Schreb.; Xu et al., 1995), perennial ryegrass (*Lolium perenne* L.; Humphreys et al., 1995a), and many others. As one examines the list of crops undergoing molecular mapping, it is apparent that little molecular mapping is occurring for forages and in particular forage grasses.

Forage grasses are generally perennial and polyploid as are about 30 to 70% of the angiosperms (Masterson, 1994). The polyploid nature of many forage grasses makes molecular mapping a challenge because of the presence of a large number of segregating genotypes, oftentimes unknown genome constitutions and chromosome pairing that is poorly understood, and complicated genotype char-


*Molecular and Cellular Technologies for Forage Improvement*. CSSA Special Publication no. 26.

acterization due to segregation and comigration of different fragments (Sorrells, 1992). In addition, forage grasses have small flowers making crossing difficult to obtain segregating progenies and most forage grasses are outcrossing species that tolerate little inbreeding, further complicating the development of mapping populations. The amount of effort going into molecular mapping of forage grasses is small in comparison to many other crops because few forage molecular biologists are present and funds available for mapping forage grasses are modest. These constraints have held back progress in development of molecular maps for forage grasses. Because most forage grasses are perennial, they can be kept almost indefinitely as clonally propagated plants, a distinct advantage compared with mapping populations of many other crop species. Vegetative propagation is particularly useful for breeders who map quantitative genes for yield or pest resistance that requires multiple-year evaluations.

The reasons for wanting a molecular map of forage grasses are similar to those of other crop species. For example, the development of molecular maps in forage grasses is expected to provide considerable insight into the organization and evolution of plant genomes. This is important because many forage grasses are complex polyploids. In addition, morphological maps are nearly absent in forage grasses and the development of molecular maps provides the best alternative in understanding the complexities of genome organization and structure. The development of molecular maps in forage grasses can provide information on assaying genetic variation within a species (Xu et al., 1994). As we look towards the future, genes for disease resistance, insect resistance, improved forage quality, and many other traits can be localized by linkage analysis, cloned on the basis of a linkage map, and transformed to produce improved cultivars. In addition, quantitative trait loci (QTLs) can be dissected, tagged, and used in marker-assisted selection to improve important traits in forage grasses. The use of molecular markers for mapping has many potential uses for the forage breeder and geneticist.

## MAPPING TECHNIQUES

The use of DNA probes, labeled radioactively or by fluorescent dyes, is a powerful means of establishing the position of base sequence difference or restriction fragment length polymorphisms (RFLPs) along a chromosome. The use of the polymerase chain reaction (PCR) to amplify particular DNA sequences so that they can be visible by gel electrophoresis has provided the basis for a larger number of methods for recognizing DNA polymorphisms between individuals, e.g., random amplified polymorphic DNA ( RAPD), variable number tandom repeats (VNTR), and amplified fragment length polymorphism (AFLP). A brief discussion of the available DNA makers for use in mapping forage grasses follows.

### Restriction Fragment Length Polymorphism

A major breakthrough for genetic mapping occurred when it was realized that genetic maps could be constructed by using pieces of chromosomal DNA as direct markers for the segregation patterns of chromosome segments (Bostein et al., 1980). Restriction fragment length polymorphisms have proven to be abundant in most organisms, and a virtually unlimited number can be mapped in any one cross. Thus it is not necessary to laboriously construct marker stocks for classical genetic mapping. Restriction fragment length polymorphisms are stable because they exhibit Mendelian codominant inheritance with minimal pleiotropic effects. In addition, DNA for RFLP mapping can be isolated from all living plant tissues at all developmental stages. They are, therefore, abundant, stable, and convenient to use in genetic studies (Helentjaris, 1987; Beckmann & Soller, 1986).

Generally RFLP analysis consists of genomic DNA isolation from a suitable set of plants, digestion of the DNA with a restriction enzyme previously screened, separation of the restriction fragments by agarose gel electrophoresis, transfer of the digested restriction fragments to a filter by Southern blotting, detection of the individual restriction fragments by DNA–DNA hybridization with a radioactively or chemically labeled cloned probe, scoring segregation by direct observation of autoradiographs, and finally map construction, e.g., using the MapMaker V2.0 or V3.0 computational programs (Lander et al., 1987).

### Random Amplified Polymorphic DNA

Random amplified polymorphic DNA markers have been widely used for genetic mapping because they provide a quick and simple method to gather information on genetic variability for a wide range of organisms. In addition the analyses can be performed with only a small amount of DNA. Random amplified polymorphic DNA markers have been used to construct genetic maps in many crops (Williams et al., 1990; Yang & Quiros, 1995). Basically, PCR amplifies DNA fragments from whole genomes by random priming. The fragments can be separated on a gel. Finally, each dominant segregating locus (band) is scored and used for map construction; however, RAPD markers have several limitations which offset some of their undoubted advantages, such as dominant-recessive characters and higher cost.

### Variable Number Tandem Repeats

One way to increase the utility of PCR based markers is to produce primers that flank genomic regions more likely to show variability than a randomly selected sequence because PCR with specific primers can only reveal polymorphisms that lie in the amplified area between the primers. Such hypervariable regions have been described in several plant species (Aswidinnoor et al., 1991; Arens et al., 1995). These regions consist of tandemly repeated DNA sequences, and allelic variability is a result of different copy numbers of the tandem repeat being present at different alleles of the same locus. The VNTR includes two types of loci termed *minisatellites* and *microsatellites*. The PCR primers flanking a VNTR locus would be expected to produce maximum length variability and to be the most useful as genetic markers.

### Amplified Fragment Length Polymorphisms

Amplified fragment length polymorphisms are a DNA fingerprinting technique resembling the RFLP technique, except that PCR amplification instead of Southern hybridization is used to detect restriction fragments. The resemblance with the RFLP technique was the basis to chose the name AFLP. The name AFLP, however, is a misnomer, because the technique will display presence or absence of particular restriction fragments rather than length differences among fragments (Reiter et al., 1992).

The AFLP technique is based on the selective PCR amplification of restriction fragments from a total digest of genomic DNA. The techniques involve three steps: (i) restriction of the DNA and ligation of oligonucleotide adaptors, (ii) selective amplification of a set of restriction fragments, and (iii) gel analysis of the amplified fragments. The selective amplification is achieved by the use of primers that extend into the restriction fragments, amplifying only those fragments in which the primer extension match the nucleotides flanking the restriction sites. Using this method, sets of restriction fragments could be visualized by PCR without knowledge of nucleotide sequence. The method allows the specific coamplification of high numbers of restriction fragments. The AFLP technique, therefore, provides a novel and very powerful DNA fingerprinting technique for any plant species.

## PARENTAL SELECTION AND MAPPING POPULATION

### Choice of Segregating Population

The choice of a suitable mapping population for the construction of molecular maps in plants will depend on the breeding system of the particular plant. So far, most RFLP maps have been produced using $F_2$ or backcross populations derived from crosses between inbred or outbred parent lines. A few maps have been constructed in forages using the $F_1$ population. Using inbred parents simplifies genetic analysis because the phase (coupling and repulsion) of markers is completely known. $F_2$ populations provide more mapping information for a given number of plants when codominant markers are analyzed, because two recombinant chromosomes can be present in each plant (Allard, 1956; Tanksley et al., 1988). $F_2$ populations provide a sex-averaged map because chromosomes from both the male and female parents are scored. Backcross populations can provide a male or female map depending on which sex was the recurrent parent. Male and female genetic maps may vary in length (De Vicente & Tanksley, 1991).

Backcross and $F_2$ populations constructed from inbred lines are segregating populations and are not permanent resources in most plants; however, most perennial forage grasses [e.g., tall fescue (Xu et al., 1995) and perennial ryegrass] can be reproduced asexually to constitute a permanent mapping population.

Mapping populations that breed true can be constructed in some plants by single-seed-descent with selfing of individual $F_2$s or backcross plants. Six or more generations of selfing are required to make plants homozygous at most loci.

The resultant recombinant inbred lines will have become largely homozygous and can be propagated by seeds. Homozygous mapping populations similar to resultant recombinant inbred lines can be produced in one step in those plants where doubled haploids can be synthesized by chromosome doubling of an $F_1$ (Powell et al., 1992). Only one meiosis has occurred, however, so there is no increase in resolution over using a $F_2$ population, but the population will breed true and will be permanent.

### Population Size

After an appropriate mapping population has been chosen, the suitable population size must be determined because the resolution of a map and the ability to determine marker order is largely dependent on population size. Clearly, population size is limited by the number of $F_2$ seeds, the number of DNA samples capably prepared in a laboratory, and cost. The larger the mapping population, the better information the map will provide. Populations with approximately 50 individuals provide less mapping resolution than larger ones but are suitable for low density maps. Moreover, if the goal is high resolution mapping in a specific region of a chromosome or linkage group or for mapping QTLs, much larger populations will be required.

Most forage grasses are outbreeding species that suffer severe inbreeding depression. Large proportions of DNA markers showing distorted segregation have been found in several grass mapping projects. For example, 27 and 37% of the RFLP markers displayed segregation distortion in a populations of 105 $F_2$ plants in tall fescue (Xu et al., 1995) and of 56 $F_2$ plants in meadow fescue (*F. pratensis* Huds.), respectively (Chen et al., 1998). Our results showed that the distorted codominant markers were skewed largely towards heterozygotes. Therefore, either increasing mapping population size or choosing easily selfed $F_1$s from outbreeding species may reduce the skewness and improve map resolution.

## GENOMIC RELATIONSHIPS WITHIN *FESTUCA* SECTION BOVINAE

Before discussing mapping results with tall and meadow fescue, it is necessary to provide a brief background on genomic relationships within *Festuca*. *Festuca* species have been subdivided by Hackel (1882) into six sections including, Ovinae, Bovinae, Sub-bulbosae, Variae, Scariosae, and Montanae (Jauhar, 1993). The first two sections include the broad- and fine-leaved *Festucas*, respectively, and have received the most attention by plant breeders and geneticists. We will limit our discussion to those species included in Bovinae because this section includes two species that are widely cultivated, namely, tall fescue and meadow fescue. Species in this section range in ploidy from diploid to decaploid. *Festuca pratensis* is recognized as a diploid ($2n = 2x = 14$) with the genomic designation of PP. The Bovinae section also includes a tetraploid ($2n = 4x = 28$), *F. arundinacea* var. *glaucescens* Boiss., with the genomic composition of

$G_1G_1G_2G_2$. The widely cultivated species, tall fescue, is a hexaploid ($2n = 6x = 42$) with genomes from meadow fescue (PP) and *F. arundinacea* var. *glaucescens* ($G_1G_1G_2G_2$). An octoploid ($2n = 8x = 56$), *F. arundinacea* var. *atlantigena*, has the genomes of $PPG_1G_1G_2G_2M_1M_1M_2M_2$. The $M_1M_1M_2M_2$ genomes are from the tetraploid *F. mairei* St. Yves. Two decaploids, *F. arundinacea* var. *letourneuxiana* St. Yves and var. *cirtensis* St. Yves have the genomic composition of $QQG_1G_1G_2G_2M_1M_1M_2M_2$ (Sleper, 1985; Sleper & West, 1996).

The genomic relationships reported for the various *Festuca* species is based largely on plant morphology and chromosome pairing relationships. Recent evidence using RFLPs in which DNA probes came from a *Pst*I-genomic library of tall fescue showed that tall fescue was closely related to meadow fescue (Xu & Sleper, 1994). In addition, meadow fescue and diploid perennial ryegrass were shown to be closely related in the same study. Humphreys et al. (1995b) used in situ hybridization to study phylogenetic relationships within tall fescue. They concluded that meadow fescue contributed one genome to tall fescue but not to *F. arundinacea* var. *glaucescens*.

## DISTRIBUTION OF *FESTUCAS* FROM SECTION BOVINAE

Tall fescue is found in much of Europe and North Africa (Borrill et al., 1971). Meadow fescue inhabits a similar area as that of tall fescue, including central and eastern Europe, parts of Scandinavia, and northern most parts of the Mediterranean. Morphological work conducted by Borrill (1972) suggested that *F. mairei*, *F. arundinacea* var. *glaucescens*, *F. arundinacea* var. *atlantigena*, and *F. arundinacea* var. *letourneuxiana* and *cirtensis* evolved in the Moroccan Atlas region.

Tall fescue is cultivated on 12 to 14 million ha in the USA and serves as the basic forage for beef (*Bos taurus*) production (Sleper & West, 1996). Tall fescue also is cultivated in most parts of Europe and the countries of Japan, southern Canada, Australia, New Zealand, Mexico, Columbia, Argentina, and parts of Africa.

## MOLECULAR MAPPING OF TALL FESCUE

Tall fescue plays an important role in world agriculture. Preliminary RFLP mapping studies in *Festuca* species described here, as an example, not only give us a better understanding of plant genome structure and species evolution, but also provide a powerful molecular tool for future forage grass breeding and genetic research. Specifically, molecular mapping in *Festuca* could be used to: (i) evaluate genetic diversity within the species, (ii) establish relationship of *Festuca* genomes that could be useful in mapping agronomically important genes or QTLs, and (iii) identify markers tightly linked to important traits that will facilitate gene transfer from alien species.

Xu et al. (1991) constructed a *Pst*I-genomic DNA library with 700 clones from a hexaploid tall fescue plant. Genomic DNA was extracted from a tall fescue plant and digested with *Pst*I. The library was constructed in pUC19 and *E. coli* XL-Blue (Stratagene) used as the host. These DNA probes have been useful in tall fescue investigations to develop molecular markers, particularly RFLPs.

After the development of the *Pst*I-genomic library in tall fescue (Xu et al., 1991) our immediate objective was to survey RFLPs in tall fescue and its relatives. We used three different *Festuca* species consisting of three ploidy levels, namely, meadow fescue ($2n = 2x = 14$), *F. arundinacea* var. *glaucescens* ($2n = 2x = 28$), and hexaploid tall fescue ($2n = 6x = 42$). The genomic DNA was restricted separately using the enzymes *Bam*HI, *Eco*RI, and *Hind*III. In the initial investigation, 174 DNA clones gave readable results with 36 (21%) representing repetitive sequences and 138 (79%) single or low copy number sequences. The greatest amount of polymorphism was found in the hexaploids followed by the tetraploids and diploids. Hexaploids averaged 70.8, 71.9, and 66.9 % while the tetraploids averaged 41.5, 43.2, and 40.9% and diploids averaged 29.5, 34.2, and 31.0% of the polymorphism detected by *Bam*HI, *Eco*RI, and *Hind*III, respectively. Diploids had the lowest amount of polymorphism, perhaps because less genetic variation is present in that germplasm compared with the other ploidy levels.

Thirty-seven of the 174 DNA probes only hybridized to either the diploids or the tetraploids (Xu et al., 1991). The represented genome-specific probes for either the P genome of meadow fescue and or the G genomes of the tetraploid *F. arundinacea* var. *glaucescens* (Fig. 2–1). Probes that did not hybridize to meadow fescue were considered to be specific to the $G_1$ and/or $G_2$ genomes while those that did not hybridize to *F. arundinacea* var. *glaucescens* were considered to be specific to the P genome of meadow fescue. These genome (species)-specific probes turned out to be useful in molecular mapping of hexaploid tall fescue as discussed later in this chapter.

Care must be taken in choosing the source of probes to use for mapping. We used our *Pst*I-genomic clones partially because they were readily available. Genomic libraries in themselves may not be the best choice to use in mapping

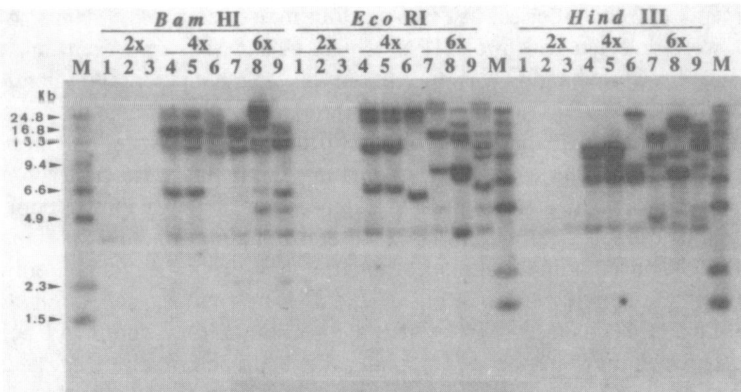


Fig. 2–1. Clone TF515 only hybridized to DNA of *F. arundinacea* var. *glaucescens* Boiss. and *F. arundinacea* var. *genuina* Schreb., indicating that *F. pratensis* Huds. lacks the homologous DNA sequence to clone TF515. Lane M is the molecular mass marker, with size marked at the left of the first lane. Genotypes 1-3, diploid (2*x*); 4-6, tetraploid (4*x*); 7-9, hexaploid (6*x*). Exposed for 12 h at –80°C.

because of their inherent inefficiencies, such as containing repeated sequences (Murray et al., 1989). Other choices, including cDNA and *Pst*I-genomic libraries, are better because they are enriched for single-copy sequences.

Our tall fescue mapping population was initiated by crossing two parents chosen carefully for their high genetic diversity so that maximum allelic differences could be expected. One parent (HD28-56) was chosen because it was used in a recurrent selection program to increase in vitro digestibility. The male parent was chosen from 'Kentucky-31,' a genetically broad-based cultivar grown throughout the USA for pasture and turf. One $F_1$ plant was selected and in an isolated greenhouse, produced 180 $F_2$ seeds out of which 105 seeds were randomly chosen to form the mapping population.

Genomic DNA clones from our *Pst*I-genomic library along with three maize probes (UMC167, CSU3, and CSU92) that identified loci on maize chromosome 1 were used. The parental and $F_2$ DNA were digested with the restriction enzymes *Bam*HI, *Eco*RI, and *Hind*III and screened with the genomic probes. MAP-MAKER (Macintosh v. 2.0; Lander et al., 1987) was used to construct the linkage map using multipoint maximum likelihood methods. The criteria for odds of marker order were at least 1000:1, and the maximum recombination ratio for linkage was 40%. Recombination frequencies were converted to centiMorgans (cM) by the Haldane mapping function (Haldane & Waddington, 1931).

### Strategy to Map Multiple Segregating Loci

Over 70% polymorphism was detected using randomly chosen *Pst*I-genomic DNA clones; however, mapping analysis appeared to be very complicated because tall fescue is polyploid and highly heterozygous, which resulted in these randomly chosen probes detecting multiple segregating fragments (Fig. 2–2); however, tall fescue is an allohexaploid that exhibits disomic inheritance (Sleper, 1985). After analysis, these multiple segregating fragments detected by single *Pst*I-genomic polymorphic probes were advantages because they allowed us to map 2 to 3 loci at once. In Fig. 2–2, four major segregating bands, A, B, C, and D, were detected with probeTF49. Bands A and D were present in only the female parent while band C was present only in the male parent with band B present in both parents. Because an $F_2$ population is inefficient in mapping dominant markers, particularly if markers are closely linked, we followed several procedures to overcome this difficulty. Segregating fragments were first scored as a dominant marker being either present or absent in the $F_2$ mapping population. Fragment B was scored as a dominant marker originating from both parents while C was scored as a dominant marker originating only from the male parent with A and D from the female parent. In Fig. 2–2, data were analyzed separately for each probe (e.g., TF49) using MAPMAKER's Two Point/Group command, followed by the Three Point/Map command. Using this approach, allelic fragments were considered to show cosegregation while nonallelic fragments located on homoeologous chromosomes were considered to be independent. Therefore, segregating fragments A and C were placed in one group and mapped at the same location because they were considered to be allelic. When fragment B originated from the male parent, it was allelic to D. When dominant B was considered as originating

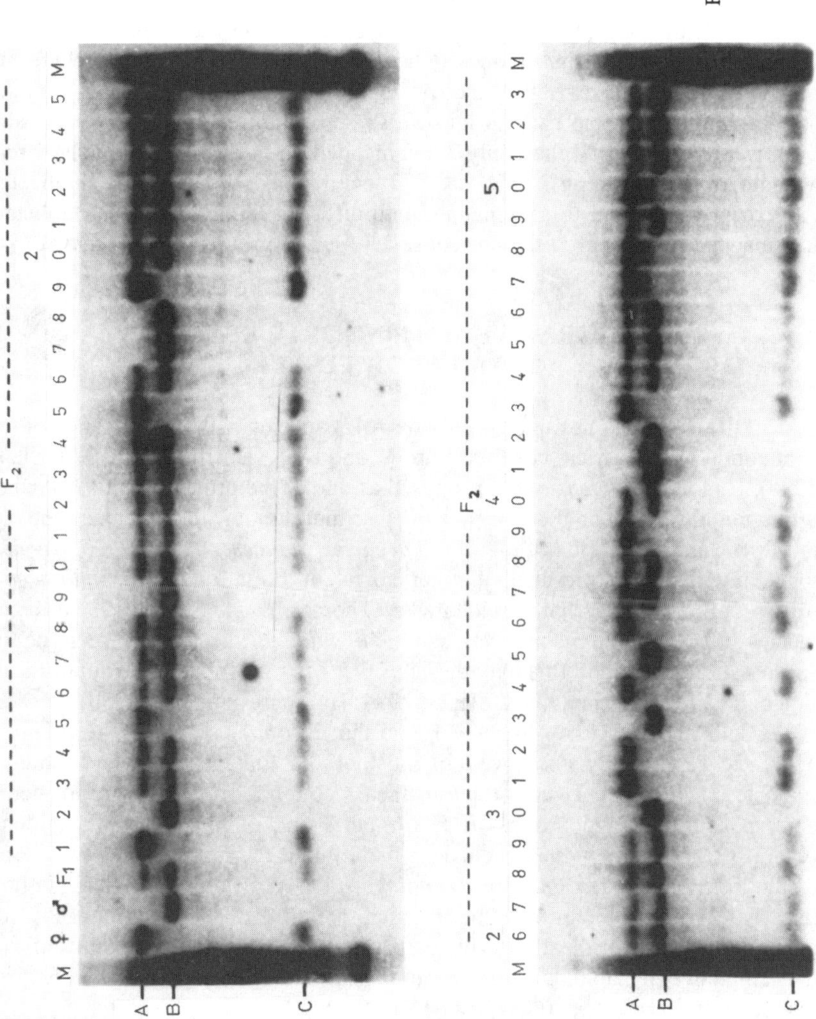


Fig. 2–2. $F_2$ segregation detected with probe TF49. Two loci are segregating, tf49AC (i.e., fragments A and C are allelic) and tf49BD. DNA was digested with *Hind*III. Results from 53 out of 105 $F_2$ plants is shown.

from the female, it was placed in its own group. The locus designation tf49AC was assigned to the segregation patterns defined by bands A and C; the locus defined by B (male parent) and D (female parent) was designated as tf49BD. Loci tf49AC and tf49BD were homoeologous and placed on different linkage groups. This procedure was used to score the segregating fragments as codominant markers. Those that did not show allelism with other fragments were treated as dominant markers.

## MOLECULAR MAPPING OF MEADOW FESCUE

Meadow fescue is a diploid ($2n = 2x = 14$) cool-season forage grass species. Despite its agricultural importance in the temperate zone, mainly in Europe (Sleper 1985), no genetic map is available due to lack of genetic markers.

An RFLP map of *F. pratensis* was generated from 56 $F_2$ plants using tall fescue *Pst*I-genomic DNA clones (Chen et al., 1998). This map included 66 markers on seven linkage groups covering a total length of 280.1 cM. On average, each linkage group had 9.4 loci with a separation of 4.2 cM between adjacent. Linkage group VII had only 2 loci mapped. Longest was the linkage group VI with 10 loci spanning 70.8 cM. Of 70 loci mapped, only 14 were duplicated. All except tf358a and tf358b mapped into different linkage groups. Duplicated loci mapped to five out of the current seven linkage groups of *F. pratensis*.

## COMPARATIVE MAPPING BETWEEN TALL AND MEADOW FESCUE

Tall fescue is a hexaploid ($2n = 6x = 42$) species with a wide range of distribution in North Africa, northern Europe, and USA (Borrill et al., 1976; Sleper 1985). Diploid *F. pratensis* was proposed as one of the progenitors of *F. arundinacea* initially based on the evidence of close morphological similarities between tall fescue and $F_1$ hybrids of *F. arundinacea* var. *glaucescens* Bioss. *F. pratensis*, together with data on presence of meiotic bivalent formation in *F. arundinacea F. pratensis* $F_1$ hybrids (Chandrasekharan & Thomas, 1971; Sleper, 1985). Recent results of RFLP comparisons and genomic *in situ* hybridization between the two species strongly support this hypothesis; *F. pratensis* is probably the donor for the P genome of tall fescue (Xu & Sleper, 1994; Humphreys et al., 1995b).

Comparative RFLP mapping found that *F. pratensis* and *F. arundinacea* genomes were highly conserved in most of the linkage groups and marker sequences (Fig. 2–3). Twenty-three of the 33 common markers were located in corresponding linkage groups in *F. pratensis* and *F. arundinacea*. Evidence obtained supports previous conclusions that diploid *F. pratensis* may be a donor of the P genome in hexaploid *F. arundinacea* based on morphological similarities, homoeologous chromosome pairing, patterns of detected RFLPs, and genomic in situ hybridization. This conclusion opens up the possibility of synthesizing novel hexaploids from diploid *F. pratensis* (PP) and tetraploid *F. arundinacea* var. *glaucescens* ($G_1G_1G_2G_2$).

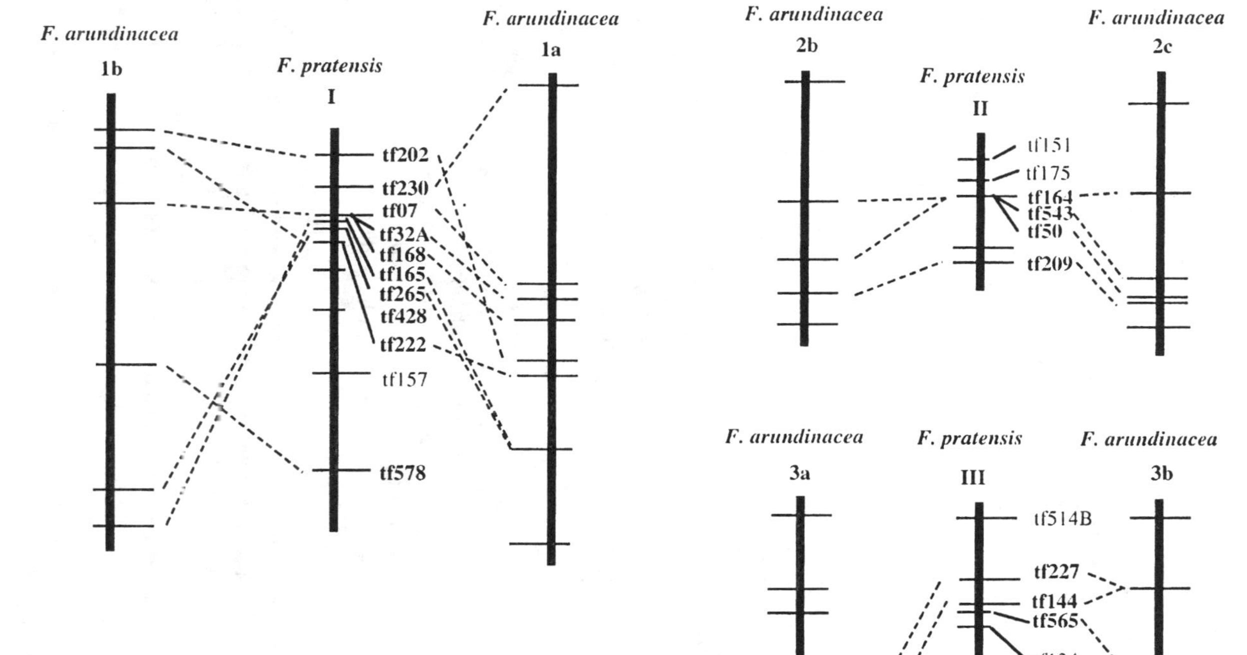


Fig. 2–3. RFLP comparative map between linkage groups I, II, and III of *F. pratensis* with corresponding linkage groups 1a, 1b, 2b, 2c, 3a, and 3b of *F. arundinacea*. Designations a, b, and c associated with *F. arundinacea* linkage groups corresponds to the three genomes comprising that species. Linkage groups are represented as thick vertical lines, with their numbering indicated at the top. Names of tall fescue RFLP markers are identified to the right of the *F. pratensis* linkage groups. The colinearity of markers between these two species is shown by dashed lines across homoeologous linkage groups.

The tetraploid $F_1$ hybrid between *F. pratensis* and *F. arundinacea* showed considerable pairing although the observed number of bivalents varied from 7 to 9.4 (Malik & Thomas, 1966); however, 14 bivalents occurred occasionally, indicating that the meadow fescue genome could pair with one of the tall fescue genomes and that the other two genomes of tall fescue paired with each other (Crowder, 1953). Later, Chandrasekhran and Thomas, (1971) tentatively proposed that *F. pratensis* and *F. arundinacea* var. *glaucescens* could be the progenitors of hexaploid *F. arundinacea* on the basis of the morphology of their $F_1$ hybrids. Xu and Sleper (1994) examined the phylogenetic relationship of tall fescue with six related species including *F. pratensis* by using RFLP data. Their results indicated that tall fescue is closely related to *F. pratensis* and tetraploid *F. arundinacea* var. *glaucescens*, and that *F. pratensis* is more closely related to perennial ryegrass than to *F. arundinacea* var. *glaucescens*.

Cytogenetic analysis based on predictive mathematical formulae to quantify relationships among genomes of different species indicated that the perennial ryegrass genome was more closely related to one of the tall fescue genomes than the other two (Kleijer, 1984). Our comparative mapping study also found that not only was there a conserved P genome between *F. pratensis* and *F. arundinacea*, but also most of the linkage groups from the P genome were closely related to those from the other two genomes, $G_1$ and $G_2$ of *F. arundinacea*. Close relationships among P, $G_1$ and $G_2$ explains why more than seven bivalents formed in 4*x* *F. pratensis F. arundinacea* hybrids.

The mapping of forage grasses is in its infancy. More work needs to be done in this area as forage grasses are of paramount importance in many parts of the world, both for livestock feed and for turf. Molecular mapping would be important in the breeding of these species for both uses. Most forage grasses are outcrossing species which makes obtaining $F_2$ seeds difficult, oftentimes requiring embryo culture techniques. It also is necessary to screen the parents to determine those which are capable of producing an abundance of $F_2$ seed because many forage grasses are self-sterile. An alternative to $F_2$ populations may be backcross progenies. Because of the highly intercrossing nature of most forage grasses, some parental materials may have abnormal chromosome segregation during meiosis so it is necessary to check chromosome pairing before harvesting $F_2$ seed. The polyploid nature of most forage grasses provides a challenge in development of comprehensive molecular maps; however, maps from polyploid species have been developed as was pointed out for tall fescue where a strategy was designed to efficiently find cosegregating loci. Development of molecular maps for forage grasses is important as we look forward to using this technology to better understand genomic relationships and to use in identifying QTLs for important agronomic traits.

## REFERENCES

Allard, R.W. 1956. Formulas and tables to facilitate the calculation of recombination values in heredity. Hilgardia 24:235–278.

Arens, P., P. Odinot, A.W. van Heusden, P. Lindhout, and B. Vosman. 1995. GATA- and GACA-repeats are not evenly distributed throughout the tomato genome. Genome 38:84–90.

Aswidinnoor, H., R.J. Nelson, J.F. Dallas, C.L. McIntyre, H. Leung, and J.P. Gustafson. 1991. Cloning and characterization of repetitive DNA sequences from genomes of *Oryza minuta* and *Oryza australiensis*. Genome 34:790–798.

Beavis, W.D., and D. Grant. 1991. A linkage map based on information from four F2 populations of maize (*Zea mays* L.). Theor. Appl. Genet. 82:636–644.

Beckmann, J.S., and M. Soller. 1986. Restriction fragment length polymorphisms and genetic improvement of agricultural species. Euphytica 35:111–124.

Berhan, A.M., S.H. Hulbert, L.G. Butler, and J.L. Bennetzen. 1993. Structure and evolution of the genomes of *Sorghum bicolor* and *Zea mays*. Theor. Appl. Genet. 86:598–604.

Borrill, M. 1972. Studies in *Festuca*. III. The contribution of the *F. scariosa* to the evolution of polyploids in sections Bovinae and Scariosae. New Phytol. 71:523–532.

Borrill, M., B. Tyler, and M. Lloyd-Jones. 1971. Studies in *Festuca*: I. A chromosome atlas of bovinae and scariosae. Cytologia 36:1–14.

Borrill, M., B.F. Tyler, and W.G. Morgan. 1976. Studies in *Festuca*: 7. Chromosome atlas (Part 2). An appraisal of chromosome race distribution ecology, including *F. pratensis* var. *apennina* (De Not.) Hacks-tetraploid. Cytologia 41:219–236.

Bostein, D., R.L. White, M. Skolnick, and R.W. Davis. 1980. Construction of a genomic linkage map in man using restriction fragment polymorphisms. J. Hum. Genet. 32:314–331.

Brummer, E.C., J.H. Bouton, and G. Kochert. 1993. Development of an RFLP map in diploid alfalfa. Theor. Appl. Genet. 86:329–332.

Chagué, V., J.C. Mercier, M. Guénard, A. de Courcel, and F. Vedel. 1996. Identification and mapping on chromosome 9 of RAPD markers linded to *Sw*-5 in tomato by bulked segregant analysis. Theor. Appl. Genet. 92:1045–1051.

Chandrasekharan, P., and H. Thomas. 1971. Studies in *Festuca*: 6. Chromosome relationships between *bovinae* and *scariosae*. Z. Pflanzenzuchtg 66:76–86.

Chen, C., D.A. Sleper, and G.S. Johal. 1998. Comparative RFLP mapping of meadow and tall fescue. Theor. Appl. Genet. 97:255–260.

Crowder, L.V. 1953. Interspecific and intergeneric hybrids of *Festuca* and *Lolium*. J Heredity 44:195–203.

De Vicente, M.C., and S.D. Tanksley. 1991. Genome-wide reduction in recombination of backcross progeny derived from male versus female gametes in an interspecific cross of tomato. Theor. Appl. Genet. 83:173–178.

Hackel, E. 1882. Monographia festucarum europearum. Theorder Fischer, Kassel, Berlin.

Haldane, J.B.S., and C.H. Waddington. 1931. Inbreeding and linkage. Genetics 16:357–374.

Hart, G.E. 1994. RFLP maps of bread wheat. p. 327–358. *In* R.L. Phillips and I.K. Vasil (ed.) DNA-based markers in plants. Advances in cellular and molecular biology of plants. Vol. I. Kluwer Academic Publ., Dordrecht, the Netherlands.

Helentjaris, T. 1987. A genetic linkage map for maize based on RFLPs. Trends Genet. 3:217–221.

Humphreys, M.O, C. Evens, M.S. Farrell, J.C. Jones, N.J. McAdam, and M.D. Hayward. 1995a. Targets for marker assisted selection in grass breeding. p. 236–237. *In* G.E. Pollott (ed.) Grassland into the 21st Century: Challenges and opportunities. Occasional Symp. 29. British Grassland Soc., Aberystwyth, Wales, England.

Humphreys, M.O., H.M. Thomas, W.G. Morgan, M.R. Meredith, J.A. Harper, H. Thomas, Z. Zwierzykowski, and M. Ghesquiére. 1995b. Discriminating the ancestral progenitors of hexaploid *Festuca arundinacea* using genomic *in situ* hybridization. Heredity 75:171–174.

Jauhar, P.P. 1993. Cytogenetics of the *Festuca–Lolium* complex, relevance to breeding. Springer-Verlag, New York.

Kilian, A., D.A. Kudrna, A. Kleinhofs, M. Yano, N. Kurata, B. Steffenson, and T. Sasaki. 1995. Rice–barley synteny and its application to saturation mapping of the barley Rpg1 region. Nucleic Acids Res. 23:2729–2733.

Kleijer, G. 1984. Cytogenetic studies of crosses between *Lolium multiflorum* Lam. and *Festuca arundinacea* Schreb: 1. The parents and the F1 hybrids. Z. Pflanzenzuchtg 93:1–22.

Kleinhofs, A., A. Kilian, M.A. Saghai Maroof, R.M. Biyashev, P. Hayes, F.G. Chen, N. Lapitan, N. Fenwick, T.K. Blake, V. Kanazin, E. Ananiev, L. Dahleen, D. Kudrna, J. Bollinger, S.J. Knapp, B. Liu, M. Sorrells, M. Heun, J.D. Franckowiak, D. Hoffman, R. Skadsen, and B.J. Steffenson. 1993. A molecular, isozyme and morphological map of the barley (*Hordeum vulgare* L.) genome. Theor. Appl. Genet. 86:705–712.

Korzun, V., G. Melz, and A. Börner. 1996. RFLP mapping of the dwarfing (*Ddw1*) and hairy peduncle (*Hp*) genes on chromosome 5 of rye (*Secale cereale* L.). Theor. Appl. Genet. 92:1073–1077.

Lander, E.S., P. Green, J. Abrahamson, A. Barlow, M.J. Daly, S.E. Lincoln, and L. Newburg. 1987. MAPMAKER: An interactive computer package for constructing primary genetic linkage maps of experimental and natural populations. Genomics 1:174–181.

Malik, C.P., and P.T. Thomas. 1966. Meiosis in the intergeneric hybrid between *Lolium multiflorum* ($2n = 14$) *Festuca arundinacea* ($2n = 70$) and its amphiploid ($2n = 84$). Z. Pflanzenzuecht 55:81–94.

Masterson, J. 1994. Stomatal size in fossil plants: evidence for polyploidy in the majority of angiosperms. Science (Washington, DC) 264:421–424.

Murray, M.G., Y. Ma, D. West, J. Romero-Severson, J. Cramer, J. Pitas, J. Kirschman., S. DeMars, and L. Vilbrandt. 1989. General considerations on building an RFLP linkage map with specific reference to maize. p. 5–9. *In* B. Burr and T. Helentjaris (ed.) Development and application of molecular markers to problems in plant genetics. Cold Spring Harbor Lab., Cold Spring Harbor, NY.

O'Donoughue, L.S., P.J. Rayapati, S.F. Kianian, M.E. Sorrells, S.D. Tanksley, M. Lee, H.W. Rines, and R.L. Phillips. 1994. Development of RFLP-based linkage maps in diploid and hexaploid oat (*Avena* sp.). p. 359–374. *In* R.L. Phillips and I.K. Vasil (ed.) DNA-based markers in plants. Advances in cellular and molecular biology of plants. Vol. I. Kluwer Academic Publ., Dordrecht, the Netherlands.

Powell, W., W.T.B. Thomas, D.M. Thompson, J.S. Swanston, and R. Waugh. 1992. Association between rDNA alleles and quantitative traits in doubled haploid populations of barley. Genetics 130:187–194.

Reiter, R.S., J.G.K. Williams, K.A. Feldmann, J.A. Rafalski, S.V. Tingey, and P.A. Scolnik. 1992. Global and local genome mapping in *Arabidosis thaliana* by using recombinant inbred lines and random amplified polymorphic DNAs. Proc. Natl. Acad. Sci. USA 89:1477–1481.

Sleper, D.A. 1985. Breeding tall fescue. Plant Breed Rev. 3:313–342.

Sleper, D.A., and C.P. West. 1996. Tall fescue. p. 471–502. *In* L.E. Moser et al. (ed.) Cool-season forage grasses. Agron. Monogr. 34. ASA, CSSA, and SSSA, Madison, WI.

Sorrells, M.E. 1992. Development and application of RFLPs in polyploids. Crop Sci. 32:1086–1091.

Tanksley, S.D., J. Miller, A. Paterson, and R. bernatzky. 1988. Molecular mapping of plant chromosomes. p. 157–173. *In* J.P. Gustafson and R. Appels (ed.) Chromosome structure and function. Plenum Publ. Corp., New York.

Webb, D.M., B.M. Baltazar, A.P. Rao-Arelli, J. Schupp, K. Clayton, P. Keim, and W.D. Beavis. 1995. Genetic mapping of soybean cyst nematode race-3 resistance loci in the soybean PI 437.654. 1995. Theor. Appl. Genet. 91:574–581.

Williams, J., A. Kubelik, K. Livak, J. Rafalski, and S. Tingy. 1990. DNA polymorphisms amplified by arbitrary primers are useful as genetic markers. Nucl. Acids Res. 18:6531–6535.

Xu, W.W., and D.A. Sleper. 1994. Phylogeny of tall fescue and related species using RFLPs. Theor. Appl. Genet. 88:685–690.

Xu, W.W., D.A. Sleper, and S. Chao. 1995. Genome mapping of polyploid tall fescue (*Festuca arundinacea* Schreb.) with RFLP markers. Theor. Appl. Genet. 91:947–955.

Xu, W.W., D.A. Sleper, and D.A. Hoisington. 1991. A survey of restriction fragment length polymorphisms in tall fescue and its relatives. Genome 34:686–692.

Xu, W.W., D.A. Sleper, and G.F. Krause. 1994. Genetic diversity of tall fescue germplasm based on RFLPs. Crop Sci. 34:246–252.

Yang, X., and C.F. Quiros. 1995. Construction of a genetic linkage map in celery using DNA-based markers. Genome 38:36–44.

# 3 Molecular Marker Applications to Genetics and Breeding of Alfalfa

**Thomas C. Osborn, Douglas J. Brouwer, K. K. Kidwell, Stefano Tavoletti, and Edwin T. Bingham**

*Department of Agronomy*
*University of Wisconsin*
*Madison, Wisconsin*

## ABSTRACT

Alfalfa (*Medicago sativa* L.) is an allogamous tetraploid with polysomic inheritance, and not the simplest system for applying molecular marker analyses; however, many advances have been made recently that provide insight into the genetics and breeding of alfalfa. Complete linkage maps of molecular markers have been developed for diploid populations, thereby avoiding the complications of tetraploid inheritance. The high levels of segregation distortion observed for inbred mapping populations can be greatly reduced by using non-inbred $F_1$ populations. Mapping data for tetraploids also has been obtained recently, and this will be important for understanding the genetic control and improving the selection of quantitative traits in populations used for cultivar development. Analysis of meiotic mutants with molecular markers has permitted mapping of centromeres and determining the mode of $2n$ gamete formation. Molecular markers also have provided information on inbreeding behavior and heterosis in alfalfa, and they may even prove useful as a tool for selecting genotypes that combine well in synthetic cultivars derived from small numbers of parents.

Molecular markers are playing an increasingly important role in genetic analysis and breeding of many crop plants. This also is true for alfalfa, although their application has been hindered by several factors. First, cultivated alfalfa is tetraploid ($2n = 4x = 32$) with polysomic inheritance. This complicates the analysis of linkage among markers and the detection of quantitative trait loci linked to markers. Fortunately, diploid genotypes ($2n = 2x = 16$) of cultivated alfalfa have been developed (Bingham & McCoy, 1979) and use of these, or the cross-fertile diploid species *M. sativa* sp. *falcata* and *M. sativa* sp. *coerulea*, greatly simplifies genetic analysis. Second, alfalfa is an allogamous species and shows severe inbreeding depression (reviewed by Jones & Bingham, 1995). This prevents the


*Molecular and Cellular Technologies for Forage Improvement*. CSSA Special Publication no. 26.

creation of highly homozygous genotypes so that most parents used in crosses have multiple alleles (up to four) at a locus and development of recombinant inbred lines for use in replicated experiments is not practical. Replication of genotypes is possible by clonal propagation, but this is labor intensive and may not be appropriate for evaluating some traits. Finally, cultivars are developed by intermating several to hundreds of selected genotypes and are released as advanced synthetic populations. This makes it difficult to integrate any form of marker assisted selection into a breeding program.

Despite these hindrances, molecular markers are providing many new insights into the genetics and breeding of alfalfa (see Brummer et al., 1994; Osborn et al., 1996 for reviews). In this chapter, we describe some of the recent insights that we have obtained using molecular markers in alfalfa. These findings are divided into two sections: (i) genetic linkage mapping, including mapping in diploids and tetraploids, and centromere mapping; and (ii) studies on inbreeding and heterosis, including the approach to homozygosity with inbreeding, the relationship between genetic distance of parents and forage yields of single-cross progenies, and the potential use of markers to select diverse parents for synthetic populations.

## GENETIC LINKAGE MAPPING

The only published molecular marker maps of alfalfa are based on segregating populations of diploid alfalfa. Use of diploid rather than tetraploid alfalfa has the obvious advantage of simplifying the linkage analysis. Diploid mapping populations segregate for only two alleles at a locus in an $F_2$ population and for only two or three alleles in a backcross population, whereas tetraploid alfalfa can segregate for up to four alleles in an $F_2$ and up to six alleles in a backcross population.

Three research groups have reported linkage maps composed of molecular markers detected as restriction fragment length polymorphisms (RFLPs) and/or random amplified polymorphic DNAs (RAPDs) in inbred populations of diploid alfalfa (Table 3–1). Kiss et al. (1993) used an $F_2$ population derived from a cross of *M. sativa* sp. *falcata* and *M. sativa* sp. *coerulea*, Brummer et al. (1993) used an $F_2$ population derived from a cross of *M. sativa* and *M. sativa* sp. *coerulea*, and Echt et al. (1994) used a backcross population derived from two *M. sativa* genotypes to develop two linkage maps, one for the $F_1$ parent and one for the recurrent parent, that were tied together by common marker loci. The maps based on inbred populations had similar total genetic distances ranging from 468 cM to 659 cM (Table 3–1), and were composed of 8 to 10 linkage groups. Common marker loci were not used in these studies, so the maps reported by the three groups cannot be aligned with each other; however, these maps are currently being expanded and integrated to provide more thorough coverage of the alfalfa genome with molecular markers.

A prominent feature of the alfalfa maps based on inbred populations is the large percentages (18–54%) of marker loci that deviated significantly from expected segregation ratios (Table 3–1). In the $F_2$ maps, the distortion tended to

Table 3–1. Characteristics of genetic linkage maps of diploid alfalfa.

| Population type | Map distance | No. and type of marker loci | Segregation distortion† | Reference |
|---|---|---|---|---|
| $F_2$ | 468 cM | 89 RFLP & RAPD | 48% | Kiss et al., 1993 |
| $F_2$ | 659 cM | 108 RFLP | 54% | Brummer et al., 1993 |
| BC (RP) | 553 cM | 61 RFLP & RAPD | 18% | Echt et al., 1994 |
| BC ($F_1$) | 603 cM | 87 RFLP & RAPD | 38% | Echt et al., 1994 |
| $F_1$ (PG-F9) | 234 cM | 50 RFLP | 7% | Tavoletti et al., 1996b |
| $F_1$ (W2x-1) | 261 cM | 55 RFLP | 11% | Tavoletti et al., 1996b |

† Percentage of all segregating marker loci showing significant ($P < 0.05$) deviation from expected ratios.

favor heterozygous genotypes. The most obvious explanation for this phenomenon is the uncovering of deleterious recessive alleles with inbreeding. This is supported by recent results from mapping in a diploid, noninbred $F_1$ population (Tavoletti et al., 1996b). In this population, only 7 and 11% of the segregating loci detected in each parent showed significant ($P < 0.05$) deviation from the expected 1:1 ratio (Table 3–1). This suggests that the use of noninbred $F_1$ populations may avoid the inaccuracies of linkage estimates due to segregation distortion in inbred populations.

Complete linkage maps are more difficult to develop for tetraploid alfalfa than for diploid alfalfa, and they have not been reported in the literature. We are working toward developing a complete linkage map using two tetraploid backcross populations. These populations were developed by crossing a nondormant, winter-sensitive genotype from the Peruvian germplasm source (PI 536535) with a dormant, winter-hardy genotype from the cultivar Blazer XL and backcrossing one $F_1$ plant to both parents. These populations segregate at RFLP loci that are heterozygous in the recurrent parent and/or the $F_1$, and the segregation patterns can be quite complex. We are developing the map using only segregation data for alleles from the $F_1$ plant and scoring only alleles that segregate as single-dose restriction fragments. Based on linkage tests, we can identify alleles that reside on the same homologue and develop four linkage maps for each chromosome (one for each homologue). The four maps can be integrated by aligning marker loci that segregate for two or more homologues.

Individual genotypes from each backcross population have been clonally propagated and evaluated for dormancy, winter survival, and freezing tolerance in replicated field trials during two winter seasons. We are using marker loci to locate genes controlling these traits to determine the numbers of genes involved, the magnitudes of gene effects, and their degrees of dominance. By comparing the map locations of loci controlling different traits we can determine if some genes affect only one trait, or obtain evidence for genes that affect multiple traits. Our preliminary analyses provide evidence that some genes affect all three traits, whereas others affect only one or two of the traits.

Another way that we have applied molecular markers to alfalfa genetics is for simultaneous mapping of centromeres and determining mode of $2n$ gamete formation in meiotic mutants. This application involves a maximum likelihood approach of half tetrad analysis (HTA) based on multiple RFLP markers, and it was tested on a diploid alfalfa clone, PG-F9, that produces $2n$ eggs at a high frequency (Tavoletti et al., 1996a). PG-F9 was pollinated with a tetraploid alfalfa and the progeny was analyzed for RFLP loci that were previously determined to be heterozygous in PG-F9 (Tavoletti et al., 1996b). Models of HTA using three linked marker loci and one unlinked marker locus were developed and tested. The results indicated the most likely position of the centromere with respect to the linked markers, and they showed that PG-F9 produced 6% first division restitution and 94% second division restitution $2n$ eggs. This method could be very useful for studying the reproductive behavior of meiotic mutants and for locating the map positions of alfalfa centromeres.

## INBREEDING AND HETEROSIS

Although alfalfa suffers from severe inbreeding depression, genotypes have been derived that are highly inbred based on the inbreeding coefficient $F$ (Posler et al., 1972; Ray & Bingham, 1992). The actual percentage of homozygous loci in inbred lines may be lower than predicted by $F$ due to inadvertent selection for heterozygosity when selecting for vigor and fertility during inbreeding. Molecular markers can be used to determine and compare the actual approach to homozygosity with the values predicted by $F$.

To test the relationship between observed and expected levels of homozygosity, we developed and analyzed inbred series from one diploid *M. sativa* genotype (MAD, four lines selfed to the $S_2$) and one diploid *M. sativa* sp. *falcata* genotype (WIF, two lines selfed to the $S_3$ and two lines selfed to the $S_4$; Brouwer & Osborn, 1997). These series were screened for fixation at 40 RFLP loci using DNA clones that detected heterozygous loci in the $S_0$ parents. The results demonstrated that for all generations of every line in both series, the observed percentage of heterozygosity was higher than expected based on $F$ (Fig. 3–1). Despite expectations as low as 6%, the lowest observed heterozygosity percentage was 30% in the WIF inbred series and 40% in the MAD inbred series. Tests of marker genotype and alleles ratios indicated that the deviations were due to selection favoring heterozygotes. These results show that the approach to homozygosity with inbreeding proceeds more slowly than expected, and that inbreeding depression may limit our ability to obtain highly homozygous lines through self-pollination.

High levels heterozygosity appear to be important for maximizing forage yields in alfalfa (Bingham, 1980), and molecular markers could be used to determine the relationship between genetic diversity of parents and yield of their hybrid progeny. We investigated this relationship by comparing the genetic distances of parents based on 61 RFLP probes to the yield of their single-cross progenies (Kidwell et al., 1994b). Four diploid ($2x$) genotypes were chromosome doubled to produce isogenic tetraploids ($4x$), and both sets of genotypes were crossed in 4 × 4 dialleles to produce six progenies at the $2x$ and $4x$ levels. Forage yields of the single-cross progenies were evaluated in spaced plant and microplot trials for 2 and 3 yr, respectively, and genetic distances (GDs) were calculated based on RFLP data for all pairwise combinations of parents at the $2x$ and $4x$ levels. The yield of diploid single-cross progenies was not significantly associated with GD; however, high correlations were detected for microplots $r = 0.69$, $P < 0.16$) and spaced plants $r = 0.82$, $P < 0.05$, Fig. 3–2) at the 4$x$ level. The higher correlations obtained for 4$x$ generations may relate to the greater proportion of heterozygotes expected at the 4$x$ level because of tetrasomic segregation and the possibility of higher order allelic interactions (Kidwell et al., 1994a; Bingham et al., 1994).

Alfalfa cultivars are developed by intermating a few to several hundred selected genotypes and then increasing the seed by open-pollination before being released as a synthetic cultivar. The positive correlation between genetic distance of parents and forage yield of single-cross progenies suggests that molecular

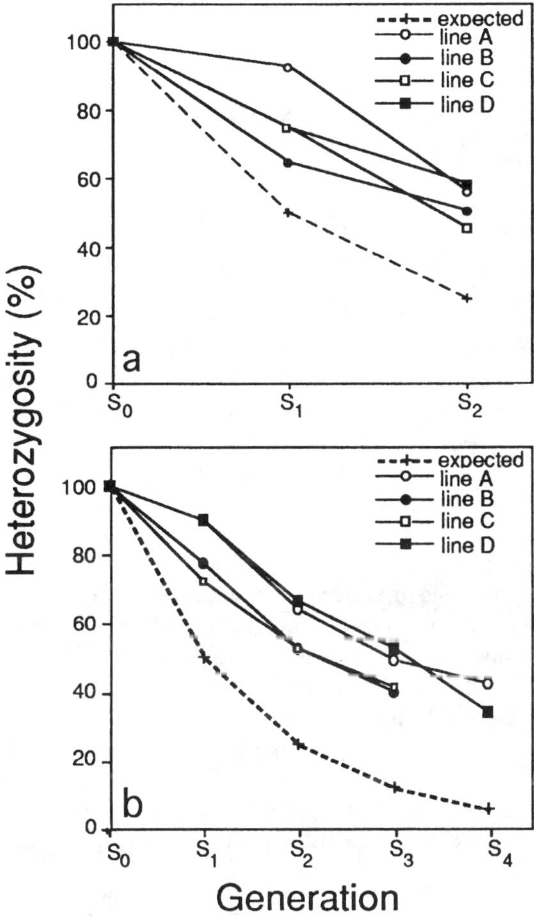


Fig. 3–1. Change in expected and observed heterozygosity with generations of inbreeding for four inbred lines from (a) MAD and four inbred lines from (b) WIF. The individual data points are for one plant ($S_0$ and $S_1$ generations of all lines, and $S_4$ generation of WIF line A) or the mean of three plants ($S_2$ and $S_3$ generations of all lines, and the $S_4$ generation of WIF line D). The expected heterozygosity equals $(1 - F)$, where $F$ is the inbreeding coefficient for each generation. (reprinted from Brouwer & Osborn, 1997).

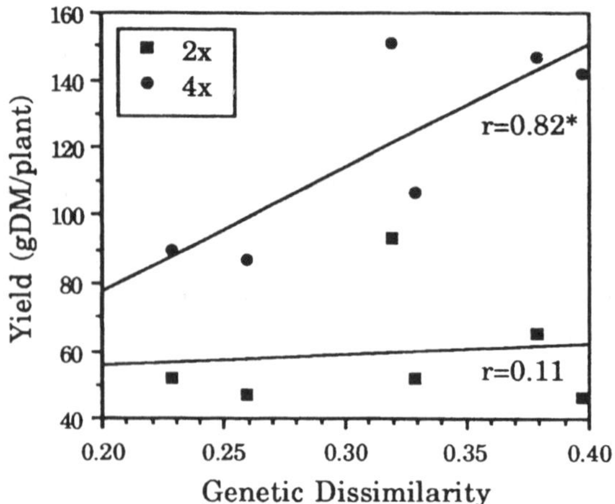


Fig. 3–2. RFLP based genetic dissimilarity vs. yield (g dry matter plant$^{-1}$) of 2*x* and 4*x* alfalfa single-cross families in spaced-plant field trial. * indicates significant *r*-value ($P > 0.05$). (reprinted from Kidwell et al., 1994b).

markers may be useful for selecting parents of high yielding synthetic cultivars. To test this approach, we screened 93 tetraploid alfalfa genotypes with 61 RFLP probes and calculated genetic distances for all 4278 possible pairs of genotypes using the compliment of the Dice coefficient as described by Kidwell et al. (1994b). Combinations of 2, 4, 8, 12, 16, and 32 parents that were either genetically similar or genetically diverse were selected by randomly choosing pairs of genotypes from the extreme 2% tails of the frequency distribution for genetic distances and then sequentially searching for the next most similar or most diverse genotype. Four samples of each parent number level were selected and the parents were intermated to produce seed for field evaluations. The resulting populations were planted in replicated trials at three locations and multiple harvests were collected for 2 yr.

The overall analysis of variance of yield data showed no significant differences among parent numbers or between genetically similar and diverse groups within parent number; however, there were significant differences among samples within genetically similar and diverse groups for some parent numbers at some locations, especially for populations with small numbers of parents. For two-thirds of the comparisons, populations made from genetically diverse parents out-yielded populations made from genetically similar parents (Kidwell et al., 1999). Further refinements of this approach may be possible to more effectively use molecular marker data for identifying parents of synthetic cultivars.

## REFERENCES

Bingham, E.T. 1980. Maximizing heterozygosity in autopolyploids. p. 471–489. *In* W.H. Lewis (ed.) Polyploidy: Biological relevance. Plenum Press, New York.

Bingham, E.T., R.W. Groose, D.R. Woodfield, and K.K. Kidwell. 1994. Complementary gene interactions in alfalfa are greater in autotetraploids than diploids. Crop Sci. 34:823–829.

Bingham, E.T., and T.J. McCoy. 1979. Cultivated alfalfa at the diploid level: Origin, reproductive stability and yield of seed and forage. Crop Sci. 19:97–100.

Brouwer, D.J., and T.C. Osborn. 1997. Molecular marker analysis of the approach to homozygosity by selfing diploid alfalfa. Crop Sci. 37:1326–1330.

Brummer, E.C., J.H. Bouton, and G. Kochert. 1993. Development of an RFLP map in diploid alfalfa. Theor. Appl. Genet. 86:329–332.

Brummer, E.C., C.S. Echt, T.J. McCoy, K.K. Kidwell, T.C. Osborn, G.B. Kiss, G. Csanadi, K. Kalman, J. Gyorgyey, L. Okresz, A.E. Raczkevy, J.H. Bouton, and G. Kochert. 1994. Molecular maps of alfalfa. p. 144–158. *In* R.L. Phillips and I.K. Vasil (ed.) DNA-based markers in plants. Kluwer Academic Publ., Dordrecht, the Netherlands.

Echt, C.S., K.K. Kidwell, S.J. Knapp, T.C. Osborn, and T.J. McCoy. 1994. Linkage mapping in diploid alfalfa *(Medicago sativa*). Genome 37:61–71.

Jones, J.S. and E.T. Bingham. 1995. Inbreeding depression in alfalfa and cross-pollinated crops. Plant Breed. Rev. 13:209–233.

Kidwell, K.K., E.T. Bingham, D.R. Woodfield, and T.C. Osborn. 1994a. Relationships among genetic distance, forage yield and heterozygosity in isogenic diploid and tetraploid alfalfa populations. Theor. Appl. Genet. 89:323–328.

Kidwell, K.K., L.M. Hartweck, B.S. Yandell, P.M. Crump, J.E. Brummer, J. Moutray, and T.C. Osborn. 1999. Forage yields of alfalfa populations derived from parents selected on the basis of molecular marker diversity. Crop Sci. (in press).

Kidwell, K.K., D.R. Woodfield, E.T. Bingham, and T.C. Osborn. 1994b. Molecular marker diversity and yield of isogenic 2*x* and 4*x* single-crosses of alfalfa. Crop Sci. 34:784–788.

Kiss, G.B., G. Csanadi, K. Kalman, P. Kalo, and L. Okresz. 1993. Construction of a basic genetic map for alfalfa using RFLP, RAPD, isozyme and morphological markers. Molec. Gen. Genet. 238:129–137.

Osborn, T.C., D.J. Brouwer, and T.J. McCoy. 1996. Molecular marker analysis of alfalfa. p. 91–109. *In* B.D. McKersie and D.C.W. Brown (ed.) Biotechnology and the improvement of forage legumes. CAB Int., Wallingford, England.

Posler, G.L., C.P. Wilsie, and R.E. Atkins. 1972. Inbreeding *Medicago sativa* L. by selfing and sib-mating, and intergenerational crossing. Crop Sci. 12:49–52.

Ray, I.M., and E.T. Bingham. 1992. Inbreeding cultivated alfalfa at the diploid level by selfing and sib-mating. Crop Sci. 32:336–339.

Tavoletti, S., E.T. Bingham, B.S. Yandell, F. Veronesi, and T.C. Osborn. 1996a. Half tetrad analysis in alfalfa using multiple RFLP markers. Proc. Natl. Acad. Sci. USA 93:10918–10922.

Tavoletti, S., F. Veronesi, and T.C. Osborn. 1996b. RFLP linkage map of an alfalfa meiotic mutant based on an F1 population. J. Hered 87:167–170.

# 4 Molecular Approaches for the Improvement of Cold Tolerance in Alfalfa[1]

**Yves Castonguay, Serge Laberge, Réal Michaud, Paul Nadeau, and Louis-P. Vézina**

*Centre de Recherches, Agriculture et Agroalimentaire Canada, Hochelaga, Sainte-Foy, Quebec, Canada*

## ABSTRACT

Lack of winterhardiness reduces the dependability of alfalfa (*Medicago sativa* L.) in cold climates. Cold tolerance is the most important factor that determines winterhardiness potential in alfalfa. A better understanding of the physiological and molecular bases of low temperature adaptation will lead to the development of new breeding strategies for the significant improvement of cold tolerance in alfalfa germplasms of high agronomic value. Changes occurring in gene expression and carbohydrate composition in crowns of cold-acclimated plants have been investigated in search of determinant traits. Electrophoretic analysis of in vitro translation products revealed extensive changes in gene expression that are very similar among cultivars of contrasting winterhardiness; however, a small number of these changes appears to be closely related to hardiness potential in alfalfa. Genes that are cold-regulated (COR) in alfalfa have been isolated and characterized. Some of these genes show a marked differential expression between cultivars of contrasting winterhardiness. Levels of galactose-containing oligosaccharides, stachyose and raffinose, were found to be better related to freezing tolerance than those of the known cryoprotectant sucrose. These results support the hypothesis that even though cold acclimation of alfalfa is a complex trait under the control of a large number of genes, differences in cold tolerance between cultivars of contrasting hardiness might be related to the differential expression of a more limited number of these genes. The potential use of COR genes and soluble sugars as molecular markers for the improvement of cold tolerance in alfalfa is currently being tested by divergent selections based on the expression of these traits within populations.

## COLD TOLERANCE OF ALFALFA AND STAND PERSISTENCE

Lack of persistence of alfalfa has historically been a major limitation to alfalfa production in areas experiencing harsh winter conditions. Costs related to alfalfa winterkill in the 1990 winter were estimated at about $200 million in

[1] Contribution no. 583 of the Centre de Recherches de Sainte-Foy.


*Molecular and Cellular Technologies for Forage Improvement.* CSSA Special Publication no. 26.

Minnesota alone (Li, 1994). Development of cultivars of high agronomic value harboring superior persistence is therefore essential to yield stability of alfalfa in cold climates and will have a significant impact on the profitability of animal productions in these regions.

Winter survival of alfalfa is determined by many factors including disease resistance, fall management, soil moisture, and cold tolerance (Beuselinck et al., 1994). Among these, low temperature tolerance was shown to have a preponderant impact on winterhardiness of alfalfa grown in northern latitudes (McKenzie et al., 1988). A close relationship between cold tolerance and adaptation of alfalfa to cold climates is illustrated by analyses of the evolution of freezing tolerance in cultivars of contrasting winterhardiness acclimated under natural winter conditions (Fig. 4–1). These studies reveal that all alfalfa cultivars have the genetic potential to increase significantly their freezing tolerance but that winter hardy cultivars consistently achieve higher levels of tolerance than non hardy cultivars. A recent evaluation of cold tolerance of alfalfa cultivars based on laboratory freezing tests in which field hardened plants sampled in fall were exposed to low subfreezing temperatures and subsequently scored for survival after regrowth were in close agreement with winter injury data from the field (Schwab et al., 1996b). Consequently, it is envisioned that improvement of freezing tolerance in germplasm of high agronomic value will translate into superior field persistence.

Successful improvement of forage legume persistence depends on sufficient genetic variability, the complexity of inheritance and the effectiveness of the procedures used to identify resistant genotypes (Smith & Kretschmer, 1989). There is a wide variability for cold tolerance within *M. sativa* (Heinrichs, 1973). Genetic diversity of many traits has been shown to be greater within than between populations of alfalfa (Heinrichs et al., 1969; Kidwell et al., 1994). According to

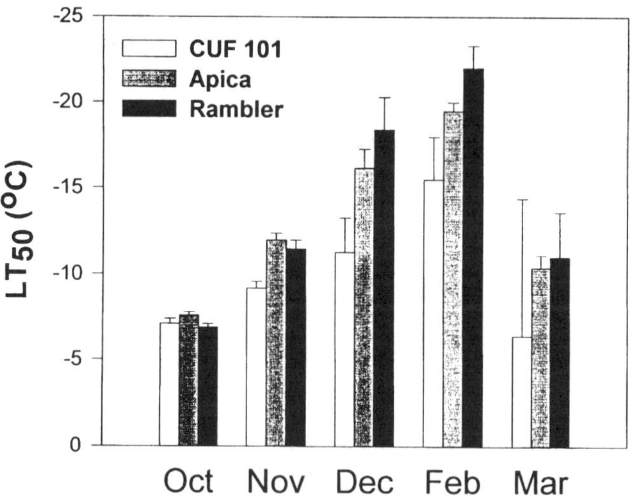


Fig. 4–1. Evolution of freezing tolerance ($LT_{50}$) of very hardy (Rambler), moderately hardy (Apica) and nonhardy (CUF 101) alfalfa cultivars acclimated to simulated natural hardening conditions in an unheated greenhouse during the 1991–1992 winter. Lower confidence limits (5%) level are presented (adapted from Castonguay et al., 1995).

Barnes et al. (1977) alfalfa is a relatively recent introduction in North America and current cultivars are derived from nine distinct sources of germplasm that differ markedly in their fall dormancy and cold hardiness potential. Cultivars with high genetic penetrance from the *M. falcata* L. germplasm, characterized by an early fall dormancy and reduced growth, are more winterhardy than those of the pure *M. sativa* background that have less fall dormancy and superior productivity (Heinrichs, 1973). The fact that early fall dormancy and reduced growth are not desirable from the productivity standpoint has prevented breeders from fully relying on very hardy sources of germplasm of the *M. falcata* type to introgress winterhardiness genes in current high-yielding alfalfa cultivars used in North America. There are, however, indications in the literature that the linkage between fall dormancy and winterhardiness is not absolute (Daday, 1964; Heichel & Henjum, 1990; McCaslin et al., 1990; Schwab et al., 1996a). Based on these observations, it should theoretically be possible to identify within the various sources of germplasms, genotypes with superior genetic potential for both yield and persistence.

## COLD HARDENING PROCESS

### Environmental Factors

Cold tolerance of alfalfa is induced in fall by exposure to low temperature, and decreasing photoperiod (McKenzie et al., 1988). Exposure to photoperiodic conditions to which the plants are adapted appears to be essential for optimal expression of the genetic potential for cold resistance in alfalfa (Hogson, 1964). According to Buxton (1989), the photoperiodic sensing mechanisms to induce the physiological changes required for cold acclimation may differ between cold-sensitive and cold-tolerant legumes. The effects of photoperiod on the induction of cold tolerance as well as its interaction with low temperature are poorly understood and deserve a closer scrutiny. Phytochrome and growth regulators have been proposed as mediators of the photoperiodic induction of fall acclimation (Williams et al., 1972; Rikin et al., 1975).

In late fall, cold hardening is mainly a function of temperature and exposure to subfreezing nonlethal temperatures is required to achieve maximum tolerance (Paquin, 1984). Similar increases in freezing tolerance at nonlethal subfreezing temperatures also have been documented in winter cereals and are described as a second stage of hardening (Livingston, 1996). When hardened under optimal conditions, hardy alfalfa can withstand subfreezing temperatures as low as −25°C for a few hours; however plants will suffer extensive frost damage when maintained for a longer period even at more moderate temperatures (−8 to −10°C) or if they are frozen in soils with high moisture content (Paquin et al., 1987). Low soil moisture and plant desiccation can promote freezing tolerance (Paquin & Méhuys, 1980; Cloutier & Andrews, 1984) by potentially triggering the expression of traits that help stabilize cellular structures during freeze-induced desiccation.

## Molecular and Genetic Bases of Cold Acclimation

More than 60 yr ago, Megee (1935) wrote: "Should it be possible to locate some one or more factors definitely correlated with winter hardiness [of alfalfa] that could be readily determined in the laboratory, this problem . . . would be greatly simplified." Although many biochemical changes occurring during cold hardening of alfalfa have been documented since then (see reviews in McKenzie et al., 1988; Buxton, 1989; Castonguay et al., 1997a), no causal relationship between the occurrence of these changes and freezing tolerance has been established thus far. A nonexhaustive list includes changes in soluble protein composition (Krasnuk et al., 1978), increased lipid unsaturation (Grenier & Willemot, 1974), as well as the accumulation of proline (Paquin, 1984), polyamines (Nadeau et al., 1987), and soluble sugars (Castonguay et al., 1995) in plants of alfalfa exposed to low temperatures. Although these changes in most cases paralleled the acquisition of freezing tolerance, large variations due to environments, the lack of differences between cultivars of contrasting hardiness or the difficulty of scaling up analytical procedures have up to now precluded the use of these biochemical traits for the improvement of cold tolerance in alfalfa.

In the past few years, our group has investigated the molecular determinants of cold adaptation in alfalfa with the aim to develop new selection approaches to significantly improve persistence in cultivars of high agronomic value. Analytical procedures such as high performance liquid chromatography, two-dimensional electrophoretic analyses of proteins, molecular biology and recombinant DNA technology in combination with conventional breeding approaches have been used to assess the genetic and molecular bases that determine cold adaptation in alfalfa. Our research efforts have been based at the outset on the hypothesis that even though cold tolerance is a trait under the control of a large number of genes, differences in cold tolerance between cultivars of contrasting winterhardiness is determined by the differential expression of a limited number of these genes. The products encoded by these genes could then be used as molecular markers to screen for cold tolerant genotypes within populations of high agronomic value. Genetic studies with potato (*Solanum comersonii* Dunn. and *S. Cardiophyllum* Lindl.; Palta & Simon, 1993) and winter field pea (*Pisum sativum* L. subsp. *arvense*; Liesenfeld et al., 1986) support the hypothesis that freezing tolerance potential could be under the control of a limited number of genes.

## Changes in Gene Expression

The concept that changes in gene expression and accumulation of specific proteins is an integral part of acclimation to environmental stress is supported by a large body of evidence (Leone et al., 1993). Earlier investigations on the biochemical changes that occur during cold acclimation of alfalfa revealed the occurrence of both quantitative and qualitative changes in protein composition (McKenzie et al., 1988). More recently, in vitro translation analyses of mRNA isolated from crowns of unhardened and hardened plants have been used to establish a potential relationship between the expression of specific cold-regulated

(COR) genes and the freezing tolerance potential in alfalfa. In a first series of experiments performed under environmentally-controlled conditions we have documented the occurrence of numerous changes in the populations of translatable mRNAs in crowns of cold acclimated plants of two alfalfa cultivars of contrasting cold adaptation (Castonguay et al., 1993). These results agreed with earlier reports by Mohapatra et al. (1987a,b, 1988) describing changes in gene expression in cold-acclimated seedlings of alfalfa. Although the two cultivars of contrasting cold adaptation gave very similar profiles under the various hardening conditions, at least three COR translation products were more highly expressed in the cold hardy Apica than in the cold sensitive CUF-101. In our study, increased freezing tolerance in plants that were incubated at a nonlethal subfreezing temperature (−2°C) as compared with plants acclimated at low nonfreezing temperatures (2°C) was associated with an enhancement of COR polypeptides and with the appearance of new translatable mRNAs. Changes in gene expression at subfreezing temperatures might be related to the increase in freezing tolerance of alfalfa in frozen soils covered with snow (Paquin, 1984).

The observation of marked changes in gene expression during the second stage of hardening at subfreezing temperatures suggests that cold acclimation studies made under environmentally-controlled conditions might not be an optimal approach to identify traits of relevance to cold adaptation under field conditions. With this consideration in mind, we have investigated the molecular bases of cold adaptation using plants of alfalfa acclimated to natural variations in temperature during fall and winter in an unheated greenhouse as described in Castonguay et al. (1995). Plants acclimated under these conditions achieve levels of freezing tolerance comparable to those observed under field conditions. This system allows convenient access to fully acclimated plant material throughout winter without the inherent difficulties associated with sampling field-grown plants in frozen soils covered with snow. Using this approach, we have recently documented that changes in gene expression observed in six cultivars of contrasting winterhardiness acclimated under simulated winter conditions were qualitatively similar to those observed in plants acclimated at nonlethal subfreezing temperatures (−2°C) under environmentally-controlled conditions (Castonguay et al., 1997b). This study confirmed the existence of a large degree of similarity in cold-induced changes in gene expression among alfalfa cultivars with few but clearly noticeable differences between cultivars of contrasting hardiness. For instance, there was a gradation in the number of peptides present in a group of low molecular weight basic polypeptides and hardiness potential in alfalfa (Fig. 4–2). In that group, three peptides were detected in the very hardy cultivars, a single polypeptide was present in the moderately hardy cultivars and none could be observed in the nonhardy cultivars. These observations on the changes in gene expression during cold acclimation of alfalfa support the hypothesis that cold tolerance potential in alfalfa might be determined by the differential expression of a limited number of COR genes. Time course analysis of the changes in gene expression in the six alfalfa cultivars revealed that the accumulation of the upregulated gene products occurred earlier in winterhardy than in nonhardy cultivars (Castonguay et al., 1997b). Variations in COR genes induction in fall could be related to differences in ABA- or $Ca^{2+}$-mediated pathways involved in low

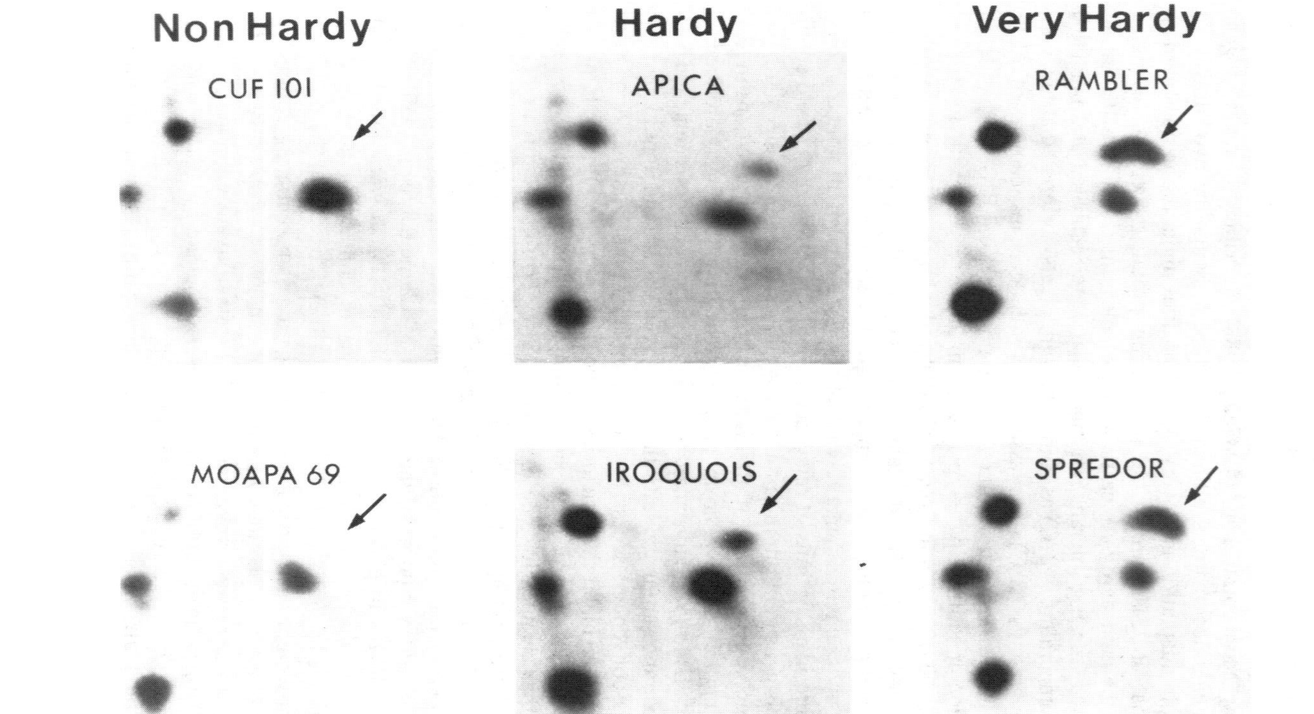


Fig. 4–2. Close-up of a low molecular weight basic area from two-dimensional polyacrylamide gels of [$^{35}$S]methionine-labeled in vitro translation products from crowns of six alfalfa cultivars from three classes of winterhardiness. Plants were acclimated to simulated winter conditions and sampled at their maximum of hardening in February. All intensely labeled polypeptides are cold induced. We note a clear gradation in the number of polypeptides in the right portion of the gels (identified by the arrow) from nonhardy to very hardy cultivars (adapted from Castonguay et al., 1997b).

Table 4–1. Structural and functional characteristics of cold-induced genes isolated by differential hybridization of a cDNA library from cold-acclimated crowns of the cultivar Apica as described in Laberge et al. (1993); ND, not determined.

| Name | Class | Transcripts size (kb) | Amino acids | Sequence information |
|---|---|---|---|---|
| msaCIA | glycine-rich | 0.9 | 204 | (GGGYNH) repeats (Laberge et al., 1993) |
| msaCIB | glycine-rich | 0.8 | 136 | Putative nuclear targeting sequence (Monroy et al., 1993a) |
| msaCIC | hybrid proline-rich | 0.9 | 166 | Pro-rich and hydrophobic domains (Castonguay et al.,1994) |
| msaCID | pathogenesis-related | 0.7 | 158 | Homologous to ABR17 (Iturriaga et al., 1994) |
| msaCIE | central metabolism | 1.2 | ND | Homologous to glyceraldehyde-3P-dehydrogenase |

temperature signal transduction (Robertson & Gusta, 1986; Monroy et al., 1993b). Low temperature stimulation of protein phosphorylation and of the accumulation of heat stable proteins in isolated nuclei of alfalfa was recently shown to be greater in the freezing-tolerant Apica than in the freezing-sensitive Trek (Kawczynski & Dhindsa, 1996).

Genes that are cold regulated in alfalfa have been isolated by differential screening of cDNA libraries constructed using mRNA extracted from cold-acclimated seedlings (Mohapatra et al., 1989) or crowns of cold-acclimated plants (Laberge et al., 1993) using single-strand cDNA probes from nonacclimated and cold-acclimated material. Using that approach, we have isolated several COR genes that in most cases encode for small polypeptides of unknown functions (Table 4–1). The cDNA clones msaCIA (Laberge et al., 1993) and msaCIB (cas15; Monroy et al., 1993a) encode putative gly-rich proteins with many repeated motifs. In vitro translation using [$^3$H]glycine as the label revealed that many COR translation products are glycine-rich (Castonguay et al., 1993). This is in contrast with very low [$^3$H]glycine labeling when mRNA from crowns of nonacclimated plants is translated in vitro. Consequently, high glycine content must have a special relevance to either the conformation, stability, or function of COR polypeptides. The msaCIC gene encodes a hybrid proline-rich protein homologous to a number of environmentally- or developmentally-regulated proteins of unknown functions (Castonguay et al., 1994). Transcripts encoded by msaCIC and one of its homologues in oilseed rape (*Brassica napus* L.; Goodwin et al., 1996) are specifically up-regulated by low temperatures and are not inducible by other environmental stimuli suggesting a direct role in the acquisition of cold tolerance. Sequence analysis of msaCID (Castonguay et al., 1996, unpublished data) revealed a very high level of amino acids correspondence with pea ABR17 pathogenesis-related protein shown to be dehydration and ABA-inducible (Iturriaga et al., 1994). The msaCIE gene (Castonguay et al., 1996, unpublished data) is homologous to glyceraldehyde-3P-dehydrogenase and might play a role in the extensive changes in soluble sugars composition that occur at low temperature. A gene homologous to glyceraldehyde-3P-dehydrogenase also was shown to be upregulated by dehydration in the resurrection plant (*Craterostigma plantagineum* Hochst; Velasco et al., 1994).

Two cold-acclimation-specific cDNAs isolated from alfalfa, cas17 (Wolfraim & Dhindsa, 1993) and cas18 (Wolfraim et al., 1993), share high sequence homologies with members of the dehydrin family of proteins. Complementary DNA clones that share sequence homology with dehydrins and proteins immunologically-related to the dehydrins have been shown to be cold-induced in many plant species (Houde et al., 1992; Neven et al., 1993; Arora & Wisniewski, 1994; Muthalif & Rowland, 1994; Salzman et al., 1996). Dehydrins are highly hydrophilic proteins of low molecular weight characterized by a high content in glycine and threonine, a conserved lysine-rich motif and a stability to boiling (Close et al., 1993). Boiling-stable proteins accumulate in response to numerous environmental and developmental changes (seed maturation, low temperature, water stress, osmotic stress, and increased salinity) that share a common dehydration component. The physiological roles of boiling-stable proteins including the dehydrins are still unknown but it has been speculated that they might prevent damage by stabilizing membranes and proteins or by solvating molecules in extensively desiccated cells (Bray, 1993).

Correlative evidence indicate that some COR gene products might be closely involved in the acquisition of freezing tolerance in alfalfa (Mohapatra et al., 1989). The analysis of the accumulation of transcripts of cold-regulated genes in two cultivars of contrasting winterhardiness showed a great variability in the expression of these genes (Castonguay et al., 1997a). We documented a superior accumulation of msaCIA and msaCIC transcripts in crowns of the cold tolerant cultivar Apica than in those of the cold-sensitive cultivar CUF 101. The fact that other COR genes from alfalfa do not display similar differential expression suggests a potential link between the expression of msaCIA and msaCIC and the acquisition of cold tolerance. Houde et al. (1992) also reported a strong correlation between the accumulation of a dehydrin-related protein (Wcs120) and levels of freezing tolerance of wheat (*Triticum aestivum* L.) genotypes of contrasting adaptation to low temperature. A close link between dehydrin accumulation and freezing tolerance is not always experimentally verified (van Zee et al., 1995; Arora & Wisniewski, 1996). Conflicting results are not unexpected considering the variety of adaptive strategies to cope with subfreezing temperatures in plants (e.g., tolerance to freeze-induced desiccation vs. supercooling). A recent report by Salzman et al. (1996) mentions that a dehydrin-like protein accumulated in desiccated grape (*Vitis labruscana* L.) buds only during the acquisition of cold tolerance. This suggests that the concept that boiling-stable proteins are involved in the development of cold tolerance through their contribution to tolerance to freeze-induced desiccation cannot be generalized and that functions specifically related to the acquisition of cold tolerance should be considered. Analyses of the interaction between COR proteins and other classes of compounds such as lipids, soluble sugars, and amino acids should provide a more comprehensive view of plant adaptation to low temperature (Gusta et al., 1996). The observation by Robertson et al. (1994) that the addition of sucrose markedly increased the capacity of boiling-stable proteins to protect protein extracts from heat denaturation is, in that respect, noteworthy.

### Adaptive Value of Soluble Sugar Accumulation

During cold acclimation plants acquire the capacity to withstand extensive cellular dehydration caused by extracellular ice formation (McKersie & Leshem, 1994). The accumulation of large amounts of nonreducing soluble sugars, mainly sucrose, as well as starch hydrolysis are commonly observed in cold acclimating tissues and are thought to be intimately related to the cold acclimation process (Kandler & Hopf, 1982). Similar changes in carbohydrate are observed during the acquisition of desiccation tolerance in seeds and pollen grains (Hoekstra & van Roekel, 1988; Horbowicz & Obendorf, 1994). In those studies, galactose-containing oligosaccharides of the raffinose series (raffinose and stachyose) were shown to be more closely related to the acquisition of desiccation tolerance of seeds than sucrose. Oligosaccharides of the raffinose family also were found to be closely related to the acquisition of freezing tolerance in many plant species probably through their stabilizing effect during freeze-induced desiccation (Hinesley et al., 1992; Stushnoff et al., 1993; Wiemken & Ineichen, 1993). In alfalfa, differences in the maximum level of freezing tolerance between nonhardy and hardy cultivars are more closely related to the capacity of the plant to accumulate stachyose and raffinose than to accumulate sucrose (Castonguay et al., 1995). Although the molecular basis of the cryoprotective action of soluble sugars is still unclear, it has been suggested that they could interact with lipid headgroups and stabilize membranes in freeze-desiccated cells (Hincha, 1990). Prevention of sucrose crystallization (Caffrey et al., 1988) and promotion of a glassy state (Koster, 1991) were also proposed as protective mechanisms conferred by these galactosyl-oligosaccharides.

## SCREENING FOR COLD TOLERANCE IN ALFALFA

### Phenotypic Selection

Even though conventional breeding methodologies have allowed progress in cold tolerance of alfalfa cultivars, the genetic complexity of the trait still constitutes a major impediment to the significant improvement of low temperature adaptation in cultivars of high agronomic value. Mass selection and recurrent phenotypic selection under harsh winter conditions are currently the most widespread approaches to identify genotypes with putative superior persistence. Plants that show a vigorous regrowth after test winters are selected and intercrossed; however screening under field conditions is a measure of winterhardiness potential as a whole and does not specifically reflect cold tolerance. The unpredictability of test winter occurrence as well as the intrinsic variability in the nature of the environmental constraints exerted on the plants from one test season to the other significantly reduce the reliability of this approach. To obviate these problems, it is often necessary to conduct trials at multiple sites during periods of many years (Limin & Fowler, 1991). Fall growth score of field-grown plants also has been used as a predictor of winterhardiness but is obviously not satisfactory considering the negative relationship between fall dormancy and alfalfa productivity (Stout & Hall, 1989). Due to the low specificity and unpredictability of the test winter approach, alternative methods are clearly needed for predicting cold tolerance potential in alfalfa. In that respect, the identification of traits that determine freezing tolerance potential will be key to the development of rapid and precise means to select hardier genotypes within alfalfa populations of high agronomic value.

Assessment of cold tolerance potential is more effectively done using methods applied under environmentally-controlled conditions (Smith & Krestchmer, 1989). Methods to evaluate cold tolerance of alfalfa have been reviewed by McKenzie et al. (1988) and Schwab et al. (1996b). These methods are based on the evaluation of plant or tissue damage after exposure to progressively declining subfreezing temperatures in controlled freezing tests. Numerous methods based on cellular leachate measurements have been developed for the rapid evaluation of plant injury. It includes measurement of electrolyte conductivity, ultraviolet light absorption, reduction of triphenyltetrazolium chloride, exosmosis of organic molecules. These laboratory methods are usually not sensitive enough to differentiate cultivars within narrow ranges of freezing tolerance (McKenzie et al., 1988). The best estimate of cold hardiness is still provided by the evaluation of whole plant recovery after regrowth.

Controlled freezing tests based on whole plant recovery have long been recognized as an effective approach to assess cold tolerance of alfalfa and as a potential means to select hardier genotypes within strains (Peltier & Tysdal, 1932). Results by Paquin and Méhuys (1980) and Schwab et al. (1996b) indicate that uniformity in testing for cold tolerance is best achieved by running these tests under standardized temperature and moisture regime. Gasser and Willemot (1974) reported that an alfalfa population obtained by intercrossing genotypes that survived the lowest temperatures in a freezing test was more freezing tolerant and had superior growth after freezing than the parental population. In pea, lines selected for their high degree of cold tolerance in an environmentally-controlled test showed superior winterhardiness under field conditions (Cousin et al., 1993). These are good indications that plants selected for cold tolerance under environmentally-controlled conditions will have a better persistence under field conditions.

## Marker-Assisted Selection

The occurrence of differential accumulations of oligosaccharides of the raffinose family and of COR gene transcripts in crowns of alfalfa cultivars of contrasting winterhardiness suggests their involvement in the determination of freezing tolerance. If this is indeed the case, these biochemical factors could serve as markers for the development of more persistent alfalfa. The requirements for a successful use of biochemical traits to screen for stress resistance have been summarized by Wery et al. (1993) and include: (i) characterization of the most important stress factor under field conditions (e.g., cold tolerance in alfalfa); (ii) identification of the mechanisms of resistance at the molecular level; (iii) development of simple, nondestructive and rapid methods to screen large numbers of genotypes even if it is an empirical estimate of the mechanisms; (iv) existence of sufficient genetic variability; (v) stability of trait expression without exceeding sensitivity to uncontrolled sources of variation (dehydration, heat, mineral deficiency, and others); and (vi) establishment of a causal link with stress tolerance rather than a mere consequence of stress exposure; comparative analysis of unrelated cultivars of contrasting adaptation is obviously not sufficient in that respect.

The link between the accumulation of RFO or COR gene products and cold tolerance is currently being verified through bidirectional selection for these traits within populations of alfalfa. A nondestructive screening approach using cold-induced accumulation of COR gene products and RFO is possible since these traits, initially characterized in crowns, also are expressed in cold-acclimated leaves and that extensive genotypic variability is found within populations (Fig. 4–3). Genotypes selected on the basis of their high or low levels of expression of

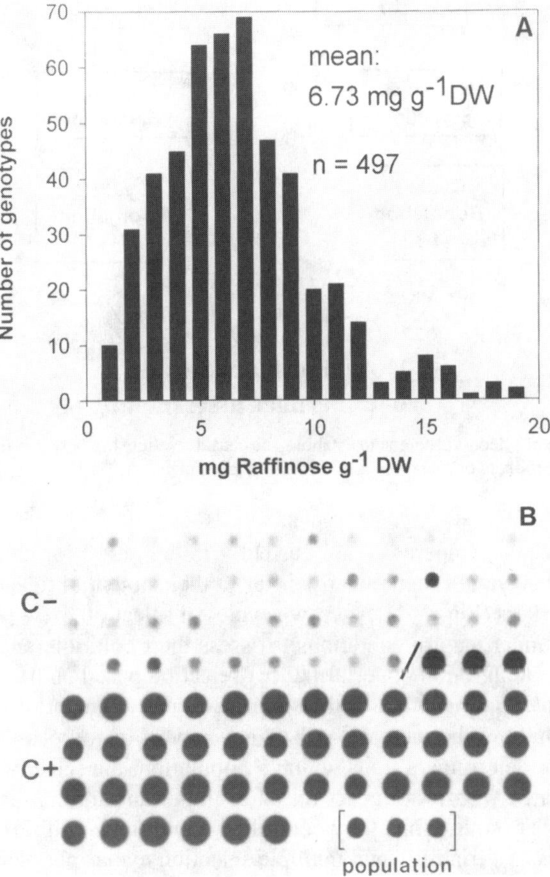


Fig. 4–3. (A) Distribution of cold-induced accumulation of leaf raffinose from nearly 500 genotypes of the cultivar DK-125. A wide distribution from nearly 0 up to 20 mg g–[1] dry weight is observed. Distribution is skewed to the right (authors, unpublished data). (B) Dot blot hybridization of a msaCIC radiolabeled cDNA probe with 10 μg total RNA from cold-acclimated leaves of individual genotypes. Extremes for low (C–) or high (C+) expression of the msaCIC gene from nearly 500 genotypes of the cultivar Ultra are presented. Dots in brackets are triplicate measurements of msaCIC expression in a pooled sample of the nearly 500 genotypes that were tested (Castonguay et al., 1996, unpublished data).

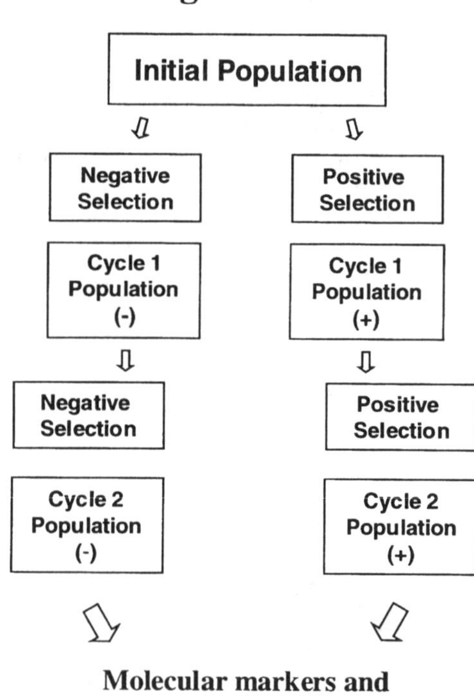


Fig. 4–4. Divergent selection scheme to establish the causal relationship between the accumulation of specific gene products or metabolites and freezing tolerance.

these traits at low temperature are currently intercrossed to generate divergent populations that should theoretically differ in their potential to accumulate these molecular markers (Fig. 4–4). After two cycles of selection these populations will be evaluated under natural conditions to assess their cold tolerance, field persistence and overall agronomic performance. Selection based on molecular markers offer many potential benefits for the significant improvement of complex-inherited traits including: the increase in the frequency of genotypes that carry favorable alleles for determinant traits within a population, the selective improvement of cold tolerance with less impact on other important agronomic traits and the screening under environmentally-controlled conditions with reduced uncontrolled sources of variability and multiple selection cycles per year.

## CONCLUDING REMARKS

Recent increases in our comprehension of the mechanisms of cold resistance at the molecular level combined with new screening technologies could lead to significant improvement of cold tolerance of alfalfa cultivars of high agronomic value. Successful improvement of legume persistence will depend on the optimal use of genetic variability and the effectiveness of procedures to identify resistant genotypes. Breeding experiments in which biochemical traits closely related to the acquisition of cold tolerance will be used as probes to select genotypes to develop populations of contrasting freezing tolerance will provide conclusive evidence on the adaptive value of these biochemical changes. Genetic engineering also could complement conventional breeding in the improvement of freezing tolerance of cold-sensitive cultivars of alfalfa. For instance, transgenic plants of alfalfa transformed with a superoxide dismutase gene from leadwort-leaved tobacco (*Nicotiana plumbaginifolia* L.) targeted to the chloroplast show superior regrowth after freezing (McKersie et al., 1993). Cosegregation of increased freezing tolerance with the superoxide dismutase transgene in the $F_1$ progeny of one transformed line supports a link between increased capacity to remove highly reactive free radicals and freezing tolerance in alfalfa. The similarities in the cold-induced changes in gene expression at low temperature in cultivars of contrasting hardiness suggest that the overexpression of genes with determinant functions in cold sensitive genotypes might lead to significant improvement in low temperature tolerance. Synchronization of the expression of transgenes, precise targeting of the gene products and their proper interaction with other cell constituents will be the key to a successful improvement of stress tolerance via genetic engineering. The availability of COR gene promoters will be essential for stress-specific expression of COR genes and to avoid imbalance that might result from their expression under unstressed conditions.

The knowledge of the molecular bases of cold tolerance will continue to markedly progress in the next few years. As a consequence, it will become increasingly feasible to specifically alter these plant characteristics through a combination of conventional and new breeding technologies. Complementary expertises in physiology, biochemistry, and plant breeding will be the key to a successful use of these techniques for the marked improvement of cold adaptation in alfalfa cultivars of high agronomic value.

## REFERENCES

Arora, R., and M. Wisniewski 1994. Cold acclimation in genetically related (sibling) deciduous and evergreen peach (*Prunus persica* [L.] Batsch): II. A 60-kilodalton bark protein in cold-acclimated tissues of peach is heat stable and related to the dehydrin family of proteins. Plant Physiol. 105:95–101.

Arora, R., and M. Wisniewski. 1996. Accumulation of a 60-kD dehydrin protein in peach xylem tissues and its relationship to cold acclimation. HortScience 31:923–925.

Barnes, D.K., E.T. Bingham, R.P. Murphy, O.J. Hunt, D.F. Beard, W.H. Skrdla, and L.R. Teuber. 1977. Alfalfa germplasm in the United States: Genetic vulnerability, use, improvement, and maintenance. USDA. Tech. Bull. 1571. USDA-ARS, Washington, DC.

Beuselinck, P.R., J.H. Bouton, W.O. Lamp, A.G. Matches, M.H. McCaslin, C.J. Nelson, L.H. Rhodes, C.C. Sheaffer, and J.J. Volenec. 1994. Improving legume persistence in forage crop systems. J. Prod. Agric. 7:311–322.

Bray, E.A. 1993. Molecular responses to water deficit. Plant Physiol. 103:1035–1040.

Buxton, D.R. 1989. Major edaphic and climatic stresses in the United States. p. 217–232. *In* G.C. Marten et al. (ed.) Persistence of forage legumes. ASA, CSSA, and SSSA, Madison, WI.

Caffrey, M., V. Fonseca, and A.C. Leopold. 1988. Lipid-sugar interactions: Relevance to anhydrous biology. Plant Physiol. 86:754–758.

Castonguay, Y., S. Laberge, P. Nadeau, and L.-P. Vézina. 1994. A cold-induced gene from *Medicago sativa* encodes a bimodular protein similar to developmentally regulated proteins. Plant Molec. Biol. 24:799–804.

Castonguay, Y., S. Laberge, P. Nadeau, and L.-P. Vézina. 1997a. Temperature and drought stress. p. 175–202. *In* B.D. McKersie and D.W. Brown (ed.). Biotechnology and the improvement of forage legumes. CAB Int., Wallingford, England.

Castonguay, Y., P. Nadeau , and S. Laberge. 1993. Freezing tolerance and alteration of translatable mRNAs in alfalfa (*Medicago sativa* L.) hardened at subzero temperatures. Plant Cell Physiol. 34:31–38.

Castonguay, Y., P. Nadeau, S. Laberge, and L.-P. Vézina. 1997b. Changes in gene expression in six alfalfa cultivars acclimated under winter hardening conditions. Crop Sci. 37:332–342.

Castonguay, Y., P. Nadeau, P. Lechasseur, and L. Chouinard. 1995. Differential accumulation of carbohydrates in alfalfa cultivars of contrasting winterhardiness. Crop Sci. 35:509–516.

Close, T.J., R.D. Fenton, A. Yang, R. Asghar, D.A. DeMason, D.E. Crone, N.C. Meyer, and F. Moonan. 1993. Dehydrin: The protein. p. 104–118. *In* T.J. Close and E.A. Bray (ed.) Plant response to cellular dehydration during environmental stress. Current topics in plant physiology. Vol 10. Am. Soc. of Plant Physiol., Rockville, MD.

Cloutier, Y., and C.J. Andrews. 1984. Efficiency of cold hardiness induction by dessication stress in four winter cereals. Plant Physiol. 767:595–598.

Cousin, R., A. Burghoffer, A. Marget, A. Vingere, and G. Eteve. 1993. Morphological, physiological and genetic bases of resistance in pea to cold and drought. p. 311–320. *In* K.B. Singh and M.C. Saxena (ed.) Breeding for stress tolerance in cool-season food legumes. John Wiley & Sons, New York.

Daday, H. 1964. Genetic relationships between cold hardiness and growth at low temperatures in *Medicago sativa*. Heredity 19:173–179.

Gasser, H., and C. Willemot. 1974. Amélioration génétique de la luzerne à la résistance au froid après une génération de sélection. Can. J. Plant Sci. 54:833–834.

Goodwin, W., J.A. Pallas, and G.I. Jenkins. 1996. Transcripts of a gene encoding a putative cell wall-plasma membrane linker proteins are specifically cold-induced in alfalfa. Plant Molec. Biol. 31:771–781.

Grenier, G., and C. Willemot. 1974. Lipid changes in roots of frost hardy and less hardy alfalfa varieties under hardening conditions. Cryobiology 11:324–331.

Gusta, L.V., R.W. Wilen, and P. Fu. 1996. Low temperature stress tolerance: The role of abscisic acid, sugars, and heat-stable proteins. HortScience 31:39–46.

Heichel, G.H., and K.I. Henjum. 1990. Fall dormancy response of alfalfa investigated with reciprocal cleft grafts. Crop Sci. 30:1123–1127.

Heinrichs, D.H., J.E. Trelsen, and F.G. Warder. 1969. Variation of chemical constituents and morphological characters within and between alfalfa populations. Can. J. Plant Sci. 49:293–305.

Heinrichs, D.H. 1973. Winterhardiness of alfalfa cultivars in southern Saskatchewan. Can. J. Plant Sci. 53:773–777.

Hincha, D.K. 1990. Differential effects of galactose containing saccharides on mechanical freeze–thaw damage to isolated thylakoid membranes. Cryo-Letters 11:437–444.

Hinesley, L.E., D.M. Pharr, L.K. Snelling, and S.R. Funderburk. 1992. Foliar raffinose and sucrose in four conifer species: Relationship to seasonal temperature. J. Am. Soc. Hortic. Sci. 117:852–855.

Hoekstra, F.A., and T. van Roekel. 1988. Desiccation tolerance of *Papaver dubium* L. pollen during its development in the anther. Plant Physiol. 88:626–632.

Hogson, H.J. 1964. Effect of photoperiod on development of cold resistance in alfalfa. Crop Sci. 4:302–305.

Horbowicz, M., and R.L. Obendorf. 1994. Seed desiccation tolerance and storability: Dependence on flatulence-producing oligosaccharides and cyclitols-review and survey. Seed Sci. Res. 4:305–405.

Houde, M., J. Danyluk, J.F. Laliberté, E. Rassart, R.S. Dhindsa, and F. Sarhan. 1992. Cloning, characterization, and expression of a cDNA encoding a 50-kilodalton protein specifically induced by cold acclimation in wheat. Plant Physiol. 99:1381–1387.

Iturriaga, E.A., M.J. Leech, D.H.P. Barratt, and T.L. Wang. 1994. Two ABA-responsive proteins from pea (*Pisum sativum* L.) are closely related to intracellular pathogenesis-related proteins. Plant Molec. Biol. 24:235–240.

Kandler, O., and H. Hopf. 1982. Oligosaccharides based on sucrose (sucrosyl oligosaccharides). p. 348–383. *In* A. Person and M.M. Zimmerman (ed.) Encyclopedia of plant physiology (New Series). Vol. 13A. Springer-Verlag, New York.

Kawczynski, W., and R.S. Dhindsa. 1996. Alfalfa nuclei contain cold-responsive phosphoproteins and accumulate heat-stable proteins during cold treatment of seedlings. Plant Cell Physiol. 37:1204–1210.

Kidwell, K.K., D.F. Austin, and T.C. Osborn. 1994. RFLP evaluation of nine *Medicago* accessions representing the original germplasm sources of North American alfalfa cultivars. Crop Sci. 34:230–236.

Krasnuk, M., F.H. Witham, and G.A. Jung. 1978. Hydrolytic enzyme differences in cold-tolerant and cold-sensitive alfalfa. Agron. J. 70:597–605.

Koster, K.L. 1991. Glass formation and desiccation tolerance in seeds. Plant Physiol. 96:302–304.

Laberge, S., Y. Castonguay, and L.-P. Vézina. 1993. New cold- and drought-regulated gene from *Medicago sativa*. Plant Physiol. 101:1411–1412.

Leone, A., R.T. Nagao, and J.J. Key. 1993. Use of molecular genetics for stress resistance breeding of cool-season food legumes. p. 323–342. *In* K.B. Singh and M.C. Saxena (ed.) Breeding for stress tolerance in cool-season food legumes. John Wiley & Sons, New York.

Li, P.H. 1994. Crop plant cold hardiness. p. 395–416. *In* K.J. Boote et al. (ed.) Physiology and determination of crop yield. ASA, CSSA, and SSSA, Madison, WI.

Liesenfeld, D.R., D.L. Auld, G.A. Murray, and J.B. Swensen. 1986. Transmittance of winter hardiness in segregated populations of peas. Crop Sci. 26:49–54.

Limin, A.E., and D.B. Fowler. 1991. Breeding for cold hardiness in winter wheat: Problems, progress and alien gene expression. Field Crops Res. 27:201–218.

Livingston, D.P., III. 1996. The second phase of cold hardening: Freezing tolerance and fructan isomer changes in winter cereal crowns. Crop Sci. 36:1568–1573.

McCaslin, M., D. Brown, H. Deery, and D. Miller. 1990. Fall dormancy and winter survival and alfalfa variation within and between populations. p. 9. *In* G.R. Bauchan et al. (ed.) Rep. of the 32nd North American Alfalfa Improvement Conf., Pasco, WA. 19–23 Aug. 1990. North Am. Alfalfa Improve., Beltsville MD.

McKenzie, J.S., R. Paquin, and S.H. Duke. 1988. Cold and heat tolerance. p. 259–302. *In* A.A. Hanson et al. (ed.) Alfalfa and alfalfa improvement. Agron. Monogr. 29. ASA, CSSA, and SSSA, Madison, WI.

McKersie, B.D., Y. Chen, M. deBeus, S.R. Bowley, C. Bowler, D. Inzé, K. D'Halluin, and J. Botterman. 1993. Superoxide dismutase enhances tolerance of freezing stress in transgenic alfalfa (*Medicago sativa* L.). Plant Physiol. 103:1150–1163.

McKersie, B.D., and Y.Y. Leshem, 1994. Stress and stress coping in cultivated plants. Academic Publ., Dordrecht, the Netherlands.

Megee, C.R. 1935. A search for factor determining winter hardiness in alfalfa. J. Am. Soc. Agron. 27:685–698.

Mohapatra, S.S., R.J. Poole, and R.S. Dhindsa. 1987a. Changes in protein patterns and translatable messenger RNA populations during cold-acclimation of alfalfa. Plant Physiol. 84:1172–1176.

Mohapatra, S.S., R.J. Poole, and R.S. Dhindsa. 1987b. Cold acclimation, freezing resistance, and protein synthesis in alfalfa (*Medicago sativa* L. cv. Saranac). J. Exp. Bot. 38:1697–1703.

Mohapatra, S.S., R.J. Poole, and R.S. Dhindsa. 1988. Abscisic acid-regulated gene expression in relation to freezing tolerance in alfalfa. Plant Physiol. 87:468–473.

Mohapatra, S.S., L. Wolfraim, and R.S. Dhindsa. 1989. Molecular cloning and relationship to freezing tolerance of cold-acclimation-specific genes of alfalfa. Plant Physiol. 89:375–380.

Monroy, A.F., Y. Castonguay, S. Laberge, F. Sarhan, L.-P. Vézina, and R.S. Dhindsa. 1993a. A new cold-induced alfalfa gene is associated with enhanced hardening at subzero temperature. Plant Physiol. 102:873–879.

Monroy, A.F., F. Sarhan, and R.S. Dhindsa. 1993b. Cold-induced changes in freezing tolerance, protein phosphorylation, and gene expression. Evidence for a role for calcium. Plant Physiol. 103:1227–1235.

Muthalif, M.M., and L.J. Rowland. 1994. Identification of dehydrin-like proteins responsive to chilling of floral buds of blueberry (*Vaccinium*, section *Cyanococcus*). Plant Physiol. 104:1439–1447.

Nadeau, P., S. Delaney, and L. Chouinard. 1987. Effects of cold hardening on the regulation of polyamine levels in wheat (*Triticum aestivum* L.) and alfalfa (*Medicago sativa* L.). Plant Physiol. 84:73–77.

Neven, L.G., D.W. Haskell, A. Hofig, Q.-B. Li, and Guy, C.L. 1993. Characterization of a spinach gene responsive to low temperature and water stress. Plant Molec. Biol. 21:291–305.

Palta, J.W., and G. Simon. 1993. Breeding potential for improvement of freezing stress resistance: Genetic separation of freezing tolerance, freezing avoidance, and capacity to cold acclimate.

p. 299–310. *In* P.H. Li and L. Christersson (ed.) Advances in cold hardiness. CRC Press, Boca Raton, FL.

Paquin, R. 1984. Influence of the environment on cold hardening and winter survival of forage plants and cereals species with consideration of proline as a metabolic marker of hardening. p. 137–154. *In* N.S. Margaris et al. (ed.) Being alive on land. Kluwer Academic Publ., Boston.

Paquin, R., M. Bernier-Cardou, and Y. Castonguay. 1987. Influence de l'humidité du sol, de la température et de la durée du gel sur la survie à l'hiver de la luzerne. Can. J. Plant Sci. 67:765–775.

Paquin, R., and G.R. Méhuys. 1980. Influence of soil moisture on cold tolerance of alfalfa. Can. J. Plant Sci. 60:139–147.

Peltier, G.L., and H.M. Tysdal. 1932. A method for the determination of comparative hardiness in seedling alfalfa by controlled hardening and artificial freezing. J. Agric. Res. 44:429–444.

Rikin, A., M. Waldman, A.E. Richmond, and A. Dovrat. 1975. Hormonal regulation of morphogenesisi and cold-resistance: I. Modifications by abscisic acid and by gibberellic acid in alfalfa (*Medicago sativa* L.) seedlings. J. Exp. Bot. 26:175–183.

Robertson, A.J., and L.V. Gusta. 1986. Abscisic acid and low temperature induced polypeptide changes in alfalfa (*Medicago sativa*) cell suspension cultures. Can. J. Bot. 64:2758–2763.

Robertson, A.J., M. Ishikawa, L.V. Gusta, and S.L. MacKenzie. 1994. Abscisic acid-induced heat tolerance in *Bromus inermis* Leyss cell-suspension cultures. Heat-stable, abscisic acid-responsive polypeptides in combination with sucrose confer enhanced thermostability. Plant Physiol. 105:181–190.

Salzman, R.A., R.A. Bressan, P.M. Hasegawa, E.N. Ashworth, and B.P. Bordelon. 1996. Programmed accumulation of LEA-like proteins during desiccation and cold acclimation of overwintering grape buds. Plant Cell Environ. 19:713–720.

Schwab, P.M., D.K. Barnes, and C.C. Sheaffer. 1996a. The relationship between field winter injury and fall growth score for 251 cultivars. Crop Sci. 36:418–426.

Schwab, P.M., D.K. Barnes, C.C. Sheaffer, and P.H. Li. 1996b. Factors affecting a laboratory evaluation of alfalfa cold tolerance. Crop Sci. 36:318–324.

Smith, R.R., and A.E. Krestschmer, Jr.. 1989. Breeding and genetics of legume persistence. p. 217–232. *In* G.C. Marten et al. (ed.) Persistence of forage legumes. ASA, CSSA, and SSSA, Madison, WI.

Stout, D.G., and J.W. Hall. 1989. Fall growth and winter survival of alfalfa in interior British Columbia. Can. J. Plant Sci. 69:491–499.

Stushnoff, C., R.L., Remmele, Jr., V. Essensee, and M. McNeil. 1993. Low temperature induced biochemical mechanisms: Implication for cold acclimation and de-acclimation. p. 647–657. *In* M.B. Jackson and C.R. Black (ed.) Interacting stresses on plants in a changing climate. NATO ASI Ser. 1. Global environmental change. Vol. 16. Springer-Verlag, New York.

van Zee, K., F. Qiang Chen, P.M. Hayes, T.J. Close, and T.H.H. Chen. 1995. Cold-specific induction of a dehydrin gene family member in barley. Plant Physiol. 108:1233–1239.

Velasco, R., F. Salamini, and D. Bartels. 1994. Dehydration and ABA increase mRNA levels and enzyme activity of cytosolic GAPDH in the resurrection plant *Craterostigma plantagineum*. Plant Molec. Biol. 26:541–546.

Wery, J., O. Turc, and J. Lecoeur. 1993. Mechanisms of resistance to cold, heat and drought in cool-season legumes, with special reference to chickpea and pea. p. 271–29. *In* K.B. Singh and M.C. Saxena (ed.) Breeding for stress tolerance in cool-season food legumes. John Wiley & Sons, New York.

Williams, B.J., Jr., N.E. Pellete, and R.M. Klein. 1972. Phytochrome control of growth cessation and initiation of cold acclimation in selected woody plants. Plant Physiol. 50:262–265.

Wiemken, V., and K. Ineichen. 1993 Effect of temperature and photoperiod on the raffinose content of spruce roots. Planta 190:387–392.

Wolfraim, L.A., and R.S. Dhindsa. 1993. Cloning and sequencing of the cDNA for *cas*17, a cold acclimation-specific gene of alfalfa. Plant Physiol. 103:667–668.

Wolfraim, L.A., R. Langis, H. Tyson, and R.S. Dhindsa. 1993. cDNA sequence, expression, and transcript stability of a cold acclimation-specific gene, *cas*18, of alfalfa (*Medicago falcata*) cells. Plant Physiol. 101:1275–1282.

# 5 Genetic Transformation of Forage Grasses

**B. V. Conger**

*Department of Plant and Soil Science*
*University of Tennessee*
*Knoxville, Tennessee*

## ABSTRACT

Genetic transformation has now been reported in numerous crops and bioengineered products are coming into the marketplace. Although this technology has not progressed as rapidly for forage and turfgrasses as it has for certain major cash crops, transgenic plants have now been obtained in several species. These include orchardgrass (*Dactylis glomerata* L.), tall fescue (*Festuca arundinacea* Schreb.), red fescue (*Festuca rubra* L.), meadow fescue (*Festuca pratensis* Huds.) perennial ryegrass (*Lolium perenne* L.), creeping bentgrass (*Agrostis palustris* Huds.), and redtop (*Agrostis alba* L.). Successful gene transfer has been accomplished by direct uptake of DNA by protoplasts and by bombardment of cells or tissues with DNA coated microprojectiles. In both cases, the transfer is followed by whole plant regeneration. Much of the work has focused on developing and improving protocols for the transformation and have used the reporter gene *uidA* coding for -glucuronidase (GUS) and the selectable marker *bar* that confers tolerance to phosphinothricin-based herbicides. Proof of the transformation has been provided by polymerase chain reaction (PCR) techniques, northern hybridization analysis of transcribed RNA, western blot analysis of soluble protein (gene product), and southern blot hybridization of total genomic DNA. Because forage grasses, in general, are not highly domesticated, possess weedy characteristics, and are highly outcrossing, special difficulties may be encountered in the ultimate release of transgenic plants.

The FLAVR SAVR tomato [*Lycopersicon lycopersicum* (L.) Karsten], introduced in 1994 by Calgene, Davis, CA (Redenbaugh et al., 1992), was the first genetically engineered food or feed crop plant to enter the marketplace. Since then several other transgenic crops have been or are being brought into commercial production and numerous others are expected during the next 5 yr (Niebling, 1995). Genes are being introduced, not only to improve quality, but also to provide tolerance to diseases, insect pests, and herbicides.

Because of the large economic investment, most of the work toward commercialization is being conducted by private companies with involvement in agricultural biotechnology. The concentration is on high value, mostly annual, cash crops. Perennial crops, such as forages, and especially grasses, receive much less


*Molecular and Cellular Technologies for Forage Improvement.* CSSA Special Publication no. 26.

attention. Among the forages, alfalfa (*Medicago sativa* L.) is the most advanced. Genetic engineering is being conducted for improved nutritional qualities (Tabe et al., 1993, 1995) and 19 permits have been issued in the USA for field testing of transgenic plants (E.T. Bingham, 1996, personal communication). Most of these are to commercial firms for genes that provide improved herbicide and pest tolerance. In addition, the University of Wisconsin is field testing alfalfa for the production of industrial enzymes, viz., Mn-dependent lignin peroxidase and alpha-amylase (Austin et al., 1995).

Spangenberg et al. (1995a) discuss the potential of improving forage grasses for improved nutritional qualities, e.g., increased quantities of sulfur-rich amino acids and reduced lignin content, pest tolerance, and others; however, most of the research conducted to date with these species has focused on developing and improving protocols for transformation and have used the reporter gene *uidA* coding for -glucuronidase, GUS, (Jefferson et al., 1987) and the selectable marker *bar* that codes for phosphinothricin acetyl transferase (PAT) and confers tolerance to the phosphinothricin based herbicides glufosinate and bialaphos and their commercial formulations Basta and Herbiace (Dennehey et al., 1994). This presentation deals with the current status regarding gene transfer in forage grasses.

## SPECIES TRANSFORMED

A complete list of forage grasses in which transgenic plants have been obtained is presented in Table 5–1. The reader also is referred to reviews by Casas et al. (1995), Jähne et al. (1995), and Lee (1996). Reports of transformed cell lines or calluses in which plants were not regenerated are not included but may be found in the reference lists of papers cited here. A few species such as, creeping bentgrass and red fescue, probably have more utility for turf than for forage; however, they are included because of the brevity of the list.

Orchardgrass was the first forage grass in which genetic transformation was demonstrated and, along with rice (*Oryza sativa* L.) and maize (*Zea mays* L.), was among the first in the family Gramineae. These reports were published in 1988 (see reviews by Conger & Kuklin, 1995; Vasil, 1995). There were no further reports of transformation in a forage grass species until 4 yr later, 1992, when transgenic plants were obtained in tall fescue by two independent groups (Ha et al., 1992; Wang et al., 1992). As shown in Table 5–1, there have been several additional reports in these and other species during the period from 1992 to 1996.

## CELLS OR TISSUES USED AND METHODS

Several methods for transferring genes to cells and obtaining transgenic plants have been described (Potrykus, 1990); however, three of these have been used predominantly and have been the most successful and repeatable. These are: (i) infection with *Agrobacterium* (*tumefaciens* or *rhizogenes*) followed by transfer of a portion of its previously genetically engineered plasmid to the recipient cell. Whole plant or shoot (followed by rooting) regeneration from the trans-

Table 5–1. Reports of forage and turf grass species in which genetically transformed plants have been obtained.

| Species | Target tissue or cells | Gene(s)† | Method‡ | Reference |
|---|---|---|---|---|
| Orchardgrass(*Dactylis glomerata* L.) | Suspension culture-derived protoplasts | *hph* | P | Horn et al., 1988b |
| | Young leaf tissue | *uidA, bar* | M | Denchev et al., 1997 |
| Tall fescue (*Festuca arundinacea* Schreb.) | Suspension culture-derived protoplasts | *hph, bar* | P | Wang et al., 1992 |
| | Suspension culture-derived protoplasts | *uidA, hph* | P | Ha et al., 1992 |
| | Suspension culture-derived protoplasts | *hph* | P | Dalton et al., 1995 |
| | Suspension culture cells | *uidA, hph, bar* | P | Spangenberg et al., 1995a |
| | Suspension culture cells | *uidA, hph* | M | Spangenberg et al., 1995b |
| Creeping bentgrass (*Agrostis palustris* Huds.) | Callus | *uidA* | M | Zhong et al., 1993 |
| | Suspension culture cells | *bar* | M | Hartman et al., 1994 |
| | Suspension culture-derived protoplasts | *bar* | P | Lee et al., 1996 |
| Redtop (*Agrostis alba* L.) | Suspension culture-derived protoplasts | *npt II* | P | Asano & Ugaki, 1994 |
| Red fescue (*Festuca rubra* L.) | Suspension culture-derived protoplasts | *bar* | P | Spangenberg et al., 1994 |
| | Suspension culture cells | *uidA, hph* | M | Spangenberg et al., 1995b |
| Perennial ryegrass (*Lolium perenne* L.) | Suspension culture cells | *uidA, hph* | M | Spangenberg et al., 1995c |
| Meadow fescue (*Festuca pratensis* Huds.) | Suspension culture-derived protoplasts | *hph, bar* | P | Spangenberg et al., 1995a |

† *hph,* Hygromycin phosphotransferase provides tolerance to the antibiotic hygromycin; *uidA*, Reporter gene for β-glucuronidase (GUS) reaction shows as blue stain when tissues are incubated in 5-bromo-4-chloro-3-indolyl—D-glucuronide (x-gluc); *bar*, Phosphinothricin acetyltransferase (PAT) provides tolerance to phosphinothricin (PPT) based herbicides such as Basta, Herbiace, bialaphos, and glufosinate; and *nptII*, Neomycin phosphotransferase.

‡ M, microprojectile bombardment; and P, direct DNA uptake using protoplasts.

formed cell is required; (ii) direct uptake of DNA (coding for the desired trait) by protoplasts followed by whole plant regeneration; and (iii) bombardment of DNA coated microprojectiles into target cells or tissues. Regeneration of a transformed plant or shoot is again required.

The *Agrobacterium* method was used to first demonstrate genetic transformation in higher plants and may still be the most widely used method for dicotyledonous species, especially for those in the family Solonaceae. It has not been a successful procedure for the Gramineae because of the apparent resistance of these species to *Agrobacterium* infection (Christou, 1995). There are, however, recent reports of successful transformation in Japonica rice (Aldemita & Hodges, 1996; Hiei et al., 1994), maize (Ishida et al., 1996), and Indica rice (Aldemita & Hodges, 1996; Rashid et al., 1996).

The first successful transformation of a forage grass (orchardgrass) used direct uptake of DNA by protoplasts. This method has been used in more than one half of the reports of transgenic forage grasses (Table 5–1). A major limitation is the difficulty in regenerating plants from protoplasts across a wide range of species and genotypes. This is a problem, not only for forage grasses, but for all gramineous species, including the major cereals (Potrykus, 1990). It has been well documented, in numerous publications, that species and genotype differences for whole plant regeneration from cells and tissues cultured in vitro exist throughout the plant kingdom. Orchardgrass was the first forage grass to be regenerated from protoplasts (Horn et al., 1988a). The genotype used, later named 'Embryogen-P' (Conger & Hanning, 1991), has a high capacity for somatic embryogenesis including the formation of embryos directly from mesophyll cells in cultured leaf segments (Conger et al., 1983). This genotype also was used in the transformation experiments.

In almost all reports of genetic transformation in forage grasses using the DNA uptake method, the protoplasts used were isolated from embryogenic suspension cultures. The uptake of DNA is facilitated by either electroporation or by adding polyethylene glycol (PEG) to the DNA protoplast mixture. Calf thymus DNA also is sometimes added as a carrier. To improve regeneration, feeder layers or nurse cultures (both use either nonmorphogenic or nonviable cells) are often included in the medium.

Invention of the gene gun or biolistic device by Sanford and colleagues at Cornell University (Klein et al., 1987) was a major advance for gene transfer technology in higher plants. For gramineous species, it has overcome the problem of no or low infection by *Agrobacterium* and the requirement of protoplast regeneration in the DNA uptake method.

The original apparatus employed a gunpowder charge to drive a plastic cylindrical macroprojectile possessing DNA coated microprojectiles (4 m tungsten particles) onto a stopping plate with a small opening (Klein et al., 1987; Sanford, 1988). The microprojectiles were released and propelled through the opening and into the tissue beyond. An improved helium-driven device using high pressure rupturable membranes rather than a gunpowder charge was described by Sanford et al. (1991). Transformation frequencies for various plant, animal, and bacterial systems ranged from 4 to 331 times higher with the helium compared with the gunpowder appliance.

Researchers at Agracetus Corporation, Middleton, WI, constructed an electrical discharge device to propel DNA-coated gold particles into tissue (Christou et al., 1988). Improvement of this methodology, including new designs of the instrument (now termed ACCELL), and its advantages over other bombardment devices, are described in a later publication (McCabe & Christou, 1993).

Development of the particle inflow gun (PIG) by Finer et al. (1992) expanded the opportunity for particle bombardment experiments to a wider range of researchers because the device is inexpensive and simple in design. It can be easily built in the laboratory from readily available parts for a few hundred dollars. Gray et al. (1994) describe the construction of a modified PIG that uses a commercially available plastic vacuum jar rather than the specially constructed steel chamber of the original. They claim that the appliance can be assembled in <40 min and that cost of the components is approximately $400. The PIG is based on acceleration of DNA-coated particles (usually tungsten) directly in a helium stream rather than being supported by a macrocarrier.

Factors such as electroporation and feeder layers were mentioned above in relation to the DNA uptake method. There also are factors and parameters to be considered in optimizing transformation using microprojectile bombardment. These have been summarized in reviews by Klein et al. (1988) and Sanford et al. (1993). McCabe et al. (1993) separate these into physical parameters, environmental factors and biological factors. A few of the most important affecting transformation of forage grasses will be described here.

Bombardment parameters studied in forage grass transformation include bombardment pressure (4–7 bar or 0.4–0.7 MPa), vacuum pressure (20–28 in Hg or ~ 500–700 mm Hg), target distance (12–20 cm), baffle mesh size (100–1000 m), distance between the baffle and target (0–10 cm), particles per bombardment (0.1–5 mg), DNA per bombardment (5–100 g), number of shots per target (1–3 shots), particle suspension volume per bombardment (5–20 L), gold or tungsten particles, particle size (1.0–3.5 m), time of preculture before bombardment, osmotic treatments before and after bombardment, and promoters driving the genes. Most of these are mentioned and discussed by Spangenberg et al. (1995b,c). Optimization of some of these will undoubtedly depend on the species and the cellular or tissue system being used.

In experiments with cultured orchardgrass leaf segments, we found that a preculture period (prior to bombardment) of 48 h produced the highest GUS expression (Denchev et al., 1995). Transformation frequencies can be increased several fold by osmotic treatments. Vain et al. (1993) propose that the mechanism is plasmolysis of the target cells such that they are less likely to extrude their protoplasm following penetration of the microprojectiles. A 30 min pretreatment of tall fescue, red fescue, and perennial ryegrass cells in liquid culture with 30 g $L^{-1}$ sucrose supplemented with 0.25 *M* mannitol and 0.25 *M* sorbitol followed by a 3 to 4 d post-bombardment period on solid medium with the same osmoticum treatment increased transformation frequencies several fold in all three species (Spangenberg et al., 1995b,c). The average number of GUS spots on orchardgrass leaf segments bombarded with DNA coated microprojectiles was increased 2.5 to 3-fold with pre- and post-treatments of 150 g $L^{-1}$ sucrose in solid medium (Denchev et al., 1997).

The cauliflower mosaic virus (CaMV) 35S promoter has been the most commonly used for transformation experiments in all plant species; however, its relative strength is much less in monocot than in dicot cells (Christensen & Quail, 1996). The maize *Adh1* promoter has been used in cereal transformation studies, but its activity appears to be restricted to root and shoot meristems, endosperm, and pollen (Kyozaka et al., 1991). The genes for rice actin, *Act1* (McElroy et al., 1990) and maize ubiquitin, *Ubi1* (Christensen et al., 1992) have been investigated as alternatives for genetic transformation of monocots. Both have been shown to be significantly more active than CaMV 35S with *Ubi1* being somewhat stronger than *Act1* (Christensen & Quail, 1996, and references therein).

Spangenberg et al. (1995b,c) used plasmids possessing either the CaMV 35S or *Act1* promoters in microprojectile bombardment experiments with tall fescue, red fescue, and perennial ryegrass; however, they did not compare the effectiveness or efficiency of the two promoters. High level GUS expression was obtained in creeping bentgrass using the *Act1* promoter (Zhong et al., 1993). The *Ubi1* promoter gave 2.3 times more blue spots (GUS expression) than the CaMV 35S promoter in microprojectile experiments with the orchardgrass leaf culture system (Denchev et al., 1995).

## GENES

Gene transfer experiments in both monocots and dicots have used the *uidA* reporter gene (GUS). Transient expression, as visualized by a blue stain, has been widely used to test the effects of different treatments and to optimize various factors and parameters, e.g., as outlined by Spangenberg et al. (1995b,c) for forage grasses. The *npt*II gene coding for neomycin phosphotransferase II and conferring tolerance to the antibiotics kanamycin, neomycin and G418 has probably been the most widely used selectable marker for obtaining stable transformation in higher plants (Fraley et al., 1986). Gramineous species, in general, are less sensitive to these antibiotics than dicotyledonous species making selection more difficult. The *hph* gene, which codes for hygromycin phosphotransferase and gives tolerance to hygromycin, is considered to be a better marker for cereals and grasses and was used in the first successful transformation with orchardgrass (Horn et al., 1988b). It has since been used in several other experiments with forage grasses (Table 5–1). As mentioned earlier, one of the most effective and popular selectable markers is the *bar* gene conferring tolerance to the phosphinothricin based herbicides. It has current wide use and selection pressure can be applied at various stages ranging from cell culture to plantlet germination to brushing leaves of mature plants with herbicide. One of the most useful constructs now in use for transformation experiments with monocots is pAHC25, which contains both the *uidA* and *bar* genes, both driven by the *Ubi1* promoter (Christensen & Quail, 1996).

As stated above, there appear to be no transgenic forage grasses with genes of agronomic importance currently in the advanced stages of testing. There are, however, field trials of the turf species, creeping bentgrass, expressing the *bar* gene (Lee et al., 1997). The potential benefit of genes conferring tolerance to disease and insect pests is obvious. Herbicide tolerance could be useful for both the establishment and maintenance of pastures. Drought and salt tolerance would allow wider adaptation and possible forage production on marginal soils or under less desirable climatic conditions. Genes to improve nutritional quality might include those coding for reduced lignin content and increasing sulfur-rich amino acids (Spangenberg et al., 1995a). Genes also are available for regulating starch biosynthesis (Stark et al., 1992). Increasing carbohydrate content could have a positive effect on forage quality.

Confirmation of genetic transformation by molecular techniques is a requirement for essentially all claims of gene transfer in higher plants. Details of these techniques are described in the original papers and therefore are mentioned only briefly here. Methods employed include: (i) amplification of polymerase chain reaction (PCR) products, sometimes followed by a Southern hybridization (Southern, 1975) of the amplified products using the coding regions of one of the inserted genes, e.g., *bar*, as a probe, (ii) northern hybridization analysis of RNA transcribed by the transgene, (iii) western blot analysis of soluble protein (gene product) encoded by the new gene, and (iv) Southern hybridization analysis of total genomic DNA from control plants and putative transformants. The latter technique is almost always performed and one or more of the others may be included as supporting evidence.

## TESTING AND RELEASE OF TRANSGENIC FORAGE GRASSES

The environmental release of transgenic crop plants, safety, and public perception has been the topic of numerous discussions and written articles. The topic is too large to cover in depth here and the reader is referred to the following recent reviews and references contained therein: Boulter (1997), Dale and Kinderlerer (1995), Gates (1995), Karieva et al. (1994), and Rogers and Parkes (1995).

A few considerations regarding genetic engineering of higher plants deserve mention. These were covered in a 1991 review (Conger, 1991) but are still pertinent today. Traditionally, most of the traits plant breeders have introduced into crops, such as dwarfing, absence of dormancy, nonshattering and uniform maturity are traits that would have an adaptive disadvantage in the wild. In contrast, many of the traits genetic engineers are interested in transferring might confer an adaptive advantage over a wild plant. These include, tolerances to drought, insects, diseases, salinity, frost, and herbicides. Also, traditional plant breeding has usually relied on crosses with closely related plants, including wild relatives, because of sexual incompatibility barriers.; however, with genetic engineering these barriers do not exist. Genes may be transferred from closely related species, distantly related species or even from organisms outside the plant kingdom such as bacteria, insects, and mammals.

The problem may not be with the genetically engineered crops themselves but rather that they may transfer these genes to wild weedy relatives through sexual hybridization. Several crop species are known to be cross-compatible with wild species as emphasized in a recent Research News Report by (Kling, 1996). Among the Gramineae, one of the potentially greatest problems is probably with sorghum [*Sorghum bicolor* (L.) Moench]. Cultivated sorghum is cross-compatible with Johnson grass (*S. halapense* L. Pers.) which is a highly noxious weed species. Annual crops that are highly domesticated, with low weediness and no weedy relatives, at least in areas where they are commonly grown, should pose less threat.

Most forage and turf grasses have undergone relatively little domestication. Hence, they possess certain weedy characteristics and are even considered as weeds in some cases. Ellstrand and Hoffman (1990) use the example of bermudagrass. This is an important turf and forage species, but it also is, in many areas of the USA, and under certain conditions, one of the worst weeds. If a gene for herbicide, salinity, or pest tolerance were to be transferred from a cultivated variety into weed populations it could have an immediate economic and ecological impact. Herbicide resistance in any forage grass species may receive considerable scrutiny before release into the environment. Therefore, regarding genetic engineering of forage grasses, breeders face not only the slow development of technology, due to lack of commercial interest and financial resources, but also the possibility of greater problems in the ultimate release and use of genetically transformed plants.

## REFERENCES

Aldemita, R.R., and T.K. Hodges. 1996. *Agrobacterium tumefaciens* -mediated transformation of *japonica* and *indica* rice varieties. Planta 199:612–617.

Asano, Y., and M. Ugaki. 1994. Transgenic plants of *Agrostis alba* obtained by electroporation-mediated direct gene transfer into protoplasts. Plant Cell Rep. 13:243–246.

Austin, S., E.T. Bingham, D.E. Matthews, M.N. Shahan, J. Will, and R.R. Burgess. 1995. Production and field performance of transgenic alfalfa (*Medicago sativa* L.) expressing alpha-amylase and manganese-dependent lignin peroxidase. Euphytica 85:381–393.

Boulter, D. 1997. Scientific and public perception of plant genetic manipulation: A critical review. Crit. Rev. Plant Sci. 16 16:231–251.

Casas, A.M., A.K. Kononowicz, R.A. Bressan, and P.M. Hasegawa. 1995. Cereal transformation through particle bombardment. p. 235–264. *In* J. Janick (ed.). Plant breeding reviews. Vol. 13. John Wiley & Sons, New York.

Christensen, A.H., and P.H. Quail. 1996. Ubiquitin promoter-based vectors for high-level expression of selectable and/or screenable marker genes in monocotyledonous plants. Transgenic Res. 5:213–218.

Christensen, A.H., R.A. Sharrok, and P.H. Quail. 1992. Maize polyubiquitin genes: Structure, thermal perturbation of expression and transcript splicing, and promoter activity following transfer to protoplasts by electroporation. Plant Molec. Biol. 18:675–689.

Christou, P. 1995. Strategies for variety-independent genetic transformation of important cereals, legumes and woody species utilizing particle bombardment. Euphytica 85:13–27.

Christou, P., D.E. McCabe, and W.F. Swain. 1988. Stable transformation of soybean callus by DNA-coated gold particles. Plant Physiol. 87:671–674.

Conger, B.V. 1991. Biotechnology in forage grass breeding. p. 147–153. *In* A.P.M. den Nijs and A. Elgersma (ed.). Fodder crops breeding: achievements, novel strategies and biotechnology. Pudoc, Wageningen, the Netherlands.

Conger, B.V., and G.E. Hanning. 1991. Registration of Embryogen-P orchardgrass germplasm with a high capacity for somatic embryogenesis from in vitro cultures. Crop Sci. 31:855.

Conger, B.V., G.E. Hanning, D.J. Gray, and J.K. McDaniel. 1983. Direct embryogenesis from mesophyll cells of orchardgrass. Science (Washington, DC) 221:850–851.

Conger, B.V., and A.I. Kuklin. 1995. In vitro culture and plant regeneration in gramineous crops. p. 59–68. *In* M. Terzi et al. (ed.). Current issues in plant molecular and cellular biology. Kluwer Acad. Publ., Dordrecht, the Netherlands.

Dale, P.J., and J. Kinderlerer. 1995. Safety in the contained use and the environmental release of transgenic crop plants. p. 36–63. *In* G.T. Tzotzos (ed.) Genetically modified organisms: A guide to biosafety. CAB Int., Oxon, England.

Dalton, S.J., A.J.E. Bettany, E. Timms, and P. Morris. 1995. The effect of selection pressure on transformation frequency and copy number in transgenic plants of tall fescue (*Festuca arundinacea* Schreb.). Plant Sci. 108:63–70.

Denchev, P.D., D. Songstad, J.K. McDaniel, and B.V. Conger. 1995. Microprojectile delivery of DNA to leaf cells in *Dactylis glomerata* and its expression in somatic embryos. p. 579–581. *In* Induced mutations and molecular techniques for crop improvement. Proc. of an FAO/IAEA Symp., Vienna, Austria. 19–23 June 1995. IAEA, Vienna.

Denchev, P.D., D.D. Songstad, J.K. McDaniel and B.V. Conger. 1997. Transgenic orchardgrass (*Dactylis glomerata*) plants by direct embryogenesis from microprojectile bombarded leaf cells. Plant Cell Rep. 16:813–819.

Dennehey, B.K., W.L. Peterson, C. Ford-Santino, M. Pajeau, and C.L. Armstrong. 1994. Comparison of selective agents for use with the selectable marker gene *bar* in maize transformation. Plant Cell Tissue Organ Cult. 36:1–7.

Ellstrand, N.C., and C.A. Hoffman. 1990. Hybridization as an avenue of escape for engineered genes. Bioscience 40:438–442.

Finer, J.J., P. Vain, M.W. Jones, and M.D. McMullen. 1992. Development of the particle inflow gun for DNA delivery to plant cells. Plant Cell Rep. 11:323–328.

Fraley, R.T., S.G. Rogers, and R.B. Horsch. 1986. Genetic transformation in higher plants. Crit. Rev. Plant Sci. 4:1–46.

Gates, P. 1995. The environmental impact of genetically engineered crops. Biotechnol. Eng. Rev. 13:181–195.

Gray, D.J., E. Hiebert, C.M. Lin, M.E. Compton, D.W. McColley, R.J. Harrison, and V.P. Gaba. 1994. Simplified construction and performance of a device for particle bombardment. Plant Cell Tissue Organ Cult. 37:179–184.

Ha, S.-B., F.-S. Wu, and T.K. Thorne. 1992. Transgenic turf-type tall fescue (*Festuca arundinacea* Schreb.) plants regenerated from protoplasts. Plant Cell Rep. 11:601– 604.

Hartman, C.L., L. Lee, P.R. Day, and N.E. Tumer. 1994. Herbicide resistant turfgrass (*Agrostic palustris* Huds.) by biolistic transformation. Bio/Technology 12:919–923.

Hiei, Y., S. Ohta, T. Komari, and T. Kumashiro. 1994. Efficient transformation of rice (*Oryza sativa* L.) mediated by *Agrobacterium* and sequence analysis of the boundaries of the T-DNA. Plant J. 6:271–282.

Horn, M.E., B. V. Conger, and C. T. Harms. 1988a. Plant regeneration from protoplasts of embryogenic suspension cultures of orchardgrass (*Dactylis glomerata* L.). Plant Cell Rep. 7:371–374.

Horn, M.E., R.D. Shillito, B.V. Conger, and C.T. Harms. 1988b. Transgenic plants of orchardgrass (*Dactylis glomerata* L.) from protoplasts. Plant Cell Rep. 7:469–472.

Ishida, Y., H. Saito, S. Ohta, Y. Hiei, T. Komari, and T. Kumashiro. 1996. High efficiency transformation of maize (*Zea mays* L.) mediated by *Agrobacterium tumefaciens*. Nature Biotechnol. 14:745–750.

Jähne, A., D. Becker, and H. Lörz. 1995. Genetic engineering of cereal crop plants: A review. Euphytica 85:35–44.

Jefferson, R.A., T.A. Kavanagh, and M.W. Bevan. 1987. GUS fusion: β Glucuronidase as a sensitive and versatile gene fusion marker in higher plants. EMBO J. 6:3901–3907.

Karieva, P., W. Morris, and C.M. Jacobi. 1994. Studying and managing the risk of cross fertilization between transgenic crops and wild relatives. Molec. Ecol. 3:15–21.

Klein, T.M., T. Gradziel, M.E. Fromm, and J.C. Sanford. 1988. Factors influencing gene delivery into *Zea mays* cells by high-velocity microprojectiles. Bio/Technology 6:559–563.

Klein, T.M., E.D. Wolf, R. Wu, and J.C. Sanford. 1987. High velocity microprojectiles for delivering nucleic acids into living cells. Nature (London) 327:70–73.

Kling, J. 1996. Could transgenic supercrops one day breed superweeds? Science (Washington, DC) 274:180–181.

Kyozuka, J., H. Fujimoto, T. Izawa, and K. Smimamoto. 1991. Anaerobic induction and tissue-specific expression of maize *Adh1* promoter in transgenic rice plants and their progeny. Molec. Gen. Genet. 228:40–48.

Lee, L. 1996. Turfgrass biotechnology. Plant Sci. 115:1–8.

Lee, L., C.L. Laramore, P.R. Day, and N.E. Tumer. 1996. Transformation and regeneration of creeping bentgrass (*Agrostis palustris* Huds.) protoplasts. Crop Sci. 36:401–406.

Lee, L., C. Laramore, C.L. Hartman, L. Yang, C.R. Funk, J. Grande, J.A. Murphy, S.A. Johnson, B.A. Majek, N.E. Tumer, and P.R. Day. 1997. Field evaluation of herbicide resistance in transgenic *Agrostis stolonifera* and inheritance in the progeny. Int. Turfgrass Soc. Res. J. 8:337–344.

McCabe, D., and P. Christou. 1993. Direct DNA transfer using electric discharge particle acceleration (ACCELL technology). Plant Cell Tissue Organ Cult. 33:227–236.

McElroy, D., W. Zhang, and R. Wu. 1990. Isolation of an efficient actin promoter for use in rice transformation. Plant Cell 2:163–171.

Niebling, K. 1995. Agricultural biotechnology companies set their sites on multi-billion $$ markets. Gen. Eng. News 15(13):1,20–21.

Potrykus, J. 1990. Gene transfer to cereals: an assessment. Bio/Technology 8:535–542.

Rashid, H., S. Yokoi, T. Toriyama, and K. Hinata. 1996. Transgenic plant production mediated by *Agrobacterium* in Indica rice. Plant Cell Rep. 15:727–730.

Redenbaugh, K., W. Hiatt, B. Martineau, M. Kramer, R. Sheehy, R. Sanders, C. Houck, and D. Emlay. 1992. Safety assessment of genetically engineered fruits and vegetables: A case study of the FLAVR SAVR tomato. CRC Press, Boca Raton, FL.

Rogers, H.J., and H.C. Parkes. 1995. Transgenic plants and the environment. J. Exp. Bot. 46:467–488.

Sanford, J.C. 1988. The biolistic process: A new concept in gene transfer and biological delivery. Trends Biotechnol. 6:229–302.

Sanford, J.C., M.J. Devit, J.A. Russell, F.D. Smith, P.R. Harpending, M.K. Roy, and S.A. Johnston. 1991. An improved helium driven biolistic device. Technique 3:3–16.

Sanford, J.C., F.D. Smith, and J.A. Russell. 1993. Optimizing the biolistic process for different biological applications. Meth. Enzymol. 217:483–509.

Southern, E.M. 1975. Detection of specific sequences among DNA fragments separated by gel electrophoresis. J. Molec. Biol. 98:503–517.

Spangenberg, G., Z.Y. Wang, J. Nagel, and I. Potrykus. 1994. Protoplast culture and generation of transgenic plants in red fescue (*Festuca rubra* L.). Plant Sci. 97:83–94.

Spangenberg, G., Z.Y. Wang, M.P. Vallés, and I. Potrykus. 1995a. Genetic transformation in *Festuca arundinacea* Schreb. (Tall Fescue) and *Festuca pratensis* Huds. (Meadow Fescue). p. 183–203. *In* Y.P.S. Bajaj (ed.). Plant protoplasts and genetic engineering. VI. Biotechnol. Agric. For. Vol. 34. Springer-Verlag, Berlin.

Spangenberg, G., Z.Y. Wang, X.L. Wu, J. Nagel, V.A. Iglesias, and I. Potrykus. 1995b. Transgenic tall fescue (*Festuca arundinacea*) and red fescue (*F. rubra*) plants from microprojectile bombardment of embryogenic suspension cells. J. Plant Physiol. 145:693–701.

Spangenberg, G., Z.Y. Wang, X. Wu, J. Nagel, and I. Potrykus. 1995c. Transgenic perennial ryegrass (*Lolium perenne*) plants from microprojectile bombardment of embryogenic suspension cells. Plant Sci. 108:209–217.

Stark, D.M., K.P. Timmerman, G.F. Berry, J. Preiss, and G.M. Kishore. 1992. Regulation of the amount of starch in plant tissues by ADP glucose pyrophosphorylase. Science (Washington, DC) 258:287–292.

Tabe, L.M., C.M. Higgins, W.C. McNabb, and T.J.V. Higgins. 1993. Genetic engineering of grain and pasture legumes for improved nutritive value. Genetica 90:181–200.

Tabe, L.M., T. Wardley-Richardson, A. Ceriotti, A. Aryan, W. McNabb, A. Moore, and T.J.V. Higgins. 1995. A biotechnological approach to improving the nutritive value of alfalfa. J. Anim. Sci. 73:2752–2759.

Vain, P., M.D. McMullen, and J.J. Finer. 1993. Osmotic treatment enhances particle bombardment-mediated transient and stable transformation of maize. Plant Cell Rep. 12:84–88.

Vasil, I.K. 1995. Cellular and molecular genetic improvement of cereals. p. 5–18. *In* M. Terzi et al. (ed.) Current issues in plant molecular and cellular biology. Kluwer Acad. Publ., Dordrecht, the Netherlands.

Wang, Z., T. Takamizo, V.A. Iglesias, M. Osusky, J. Nagel, I. Potrykus, and G. Spangenberg. Transgenic plants of tall fescue (*Festuca arundinacea* Schreb.) obtained by direct gene transfer to protoplasts. Bio/Technology 10:691–696.

Zhong, H., M.G. Bolyard, C. Srinivasan, and M.B. Sticklen. 1993. Transgenic plants of turfgrass (*Agrostis palustris* Huds.) from microprojectile bombardment of embryogenic callus. Plant Cell Rep. 13:1–6.

# 6 Molecular Analysis of Alfalfa Root Vegetative Storage Proteins[1]

**J. J. Volenec, B. C. Joern, and S. M. Cunningham**

*Department of Agronomy*
*Purdue University*
*West Lafayette, Indiana*

**A. Ourry**

*Unit Associee INRA*
*Physiologie et Biochimie Vegetales*
*Institut de Recherche en Biologie Appliquee*
*Universite de Caen, Cedex, France*

## ABSTRACT

Our long-term goal is to improve persistence and yield of alfalfa (*Medicago sativa* L.) and other forage legumes by identifying and manipulating genes that affect these traits. Future improvements by genetic manipulation depend, however, on new insights into basic physiological and biochemical plant processes. Currently we lack knowledge of discrete traits controlling agronomic performance that can serve as targets for manipulation using modern genetic techniques. Our work, and recent work of others, has failed to show a positive association between root total nonstructural carbohydrate (TNC) levels and genetic variation in regrowth and winterhardiness of forage legumes. We are exploring alternatives to the conventional thinking that root TNC reserves control alfalfa regrowth and persistence. Recent results indicate that root N declines during herbage regrowth after defoliation, and again in spring when shoot growth resumes. Labeling studies have proven that much of the N found in shoots during early regrowth is derived from root N pools. In alfalfa, certain root N pools, especially root vegetative storage proteins (VSPs) are preferentially used as N reserves during the early stages of shoot regrowth. The VSPs represent 25% of the root protein pool. They are unique to alfalfa roots, and their synthesis is developmentally regulated. Work is underway to isolate and characterize the cDNAs for the VSPs to learn more about regulation of VSP synthesis and degradation in alfalfa roots.

Forage legumes are essential components of U.S. agriculture. Alfalfa is the most important forage legume in the USA with 24 million acres harvested in 1994. Using realistic estimates for production per acre (4.0 T $yr^{-1}$) and price per ton

[1] A contribution from the Purdue University Agricultural Experiment Station, West Lafayette, IN 47907.


*Molecular and Cellular Technologies for Forage Improvement.* CSSA Special Publication no. 26.

($100), the cash value of alfalfa hay in the USA in 1994 was nearly $10 billion dollars. This does not include the value-added contribution alfalfa makes to the animal sector of the U.S. agricultural economy. In 1994 the value of all ruminant livestock, the prime users of forage like alfalfa, was approximately $60 billion dollars.

A major challenge facing forage legume production is poor plant persistence caused by environmental stress (Beuselinck et al., 1994). This problem is aggravated by the current emphasis on intensive harvest management, including late autumn harvesting. Our long-term goal is to improve persistence and yield of alfalfa and other forage legumes by identifying genes that affect agronomic performance. Plant breeding continues to be the primary mechanism by which we increase yield potential and stress tolerance of forage legumes. Future improvements by plant breeders and geneticists will depend on new insights into basic physiological and biochemical plant processes controlling yield and persistence; however, we lack knowledge of discrete mechanisms that control agronomic performance of forage legumes that can serve as targets for manipulation using modern genetic techniques. The goal of this paper is to briefly examine the broadly accepted view that root nonstructural carbohydrates control legume regrowth and persistence. We will then focus on the role of root N reserves and the growing body of evidence that suggests that amino acid and protein pools in roots can influence agronomic performance of alfalfa and other forages. A detailed review on the role of N reserves in growth and stress tolerance of forages has been published recently (Volenec et al., 1996).

## ROOT CARBOHYDRATE RESERVES

Root carbohydrate reserves are accepted by many forage researchers as the physiological factors controlling shoot growth and persistence of forage legumes. Early studies by Graber et al. (1927) were among the first to report that root nonstructural carbohydrate concentrations declined in spring in temperate environments as legumes resumed shoot growth, and again after defoliation. Numerous published studies have followed, most reaffirming the correlation between root carbohydrate reserves and persistence or forage yield of alfalfa and other perennial legumes (Grandfield, 1943; Kust & Smith, 1961; Smith, 1962, 1964; Reynolds, 1971; for reviews see Brown et al., 1972; Heichel et al., 1988).

Not all researchers agree, however, that high levels of root carbohydrates are essential for rapid shoot regrowth and/or good forage legume survival. May (1960) questioned the cause–effect relationship between root carbohydrate depletion and shoot regrowth, and suggested that root respiration may be partially responsible for the decline in root carbohydrates after harvest. These questions led to research where root constituents were labeled with $^{14}C$ prior to defoliation. The movement of $^{14}C$ from roots to regrowing shoots would verify that root starch and sugar (collectively known as total nonstructural carbohydrates, TNC) pools were used to supply C to shoots after harvest (Pearce et al., 1969; Smith & Marten, 1970). These methods are not suitable for achieving this goal, however, because nearly all organic cellular constituents in roots would be labeled with $^{14}C$ (for example, sugars, starch, *and* N-containing amino acids and proteins), making it impossible to discriminate the true source of label found in regrowing shoots. The magnitude of this problem has recently been quantified by Avice et al. (1996b) who found that up to 58% of the C transferred from roots to regrowing alfalfa shoots originated from N-containing compounds and not nonstructural carbohydrates.

A poor relationship between root TNC concentrations and shoot growth or stress tolerance of perennial legumes has been reported in several studies. Hodgson and Bula (1956) examined 11 sweetclover cultivars (*Melilotus alba* L.) and found no relationship between root TNC concentrations and cold resistance. Jung and Smith (1961) reported similar results for alfalfa and red clover (*Trifolium pratense* L.). Studies focusing on flexible autumn harvest schedules for alfalfa also have shown little relationship between root TNC levels in autumn, plant survival, and spring shoot growth. In Minnesota, Brink and Marten (1989) reduced root TNC concentrations in November by increasing harvest frequency or decreasing the interval between harvests, but the varied root TNC levels did not explain many of the differences in alfalfa persistence and spring forage yield. Similar results also have been reported for autumn management studies conducted in West Virginia (Edmisten & Wolf, 1988; Edmisten et al., 1988) and Georgia (Brown et al., 1990). Studies where cutting management is used to create different root TNC levels and subsequent plant responses monitored must be interpreted with care. It is likely that differential cutting schedules change many physiological attributes of forage legumes (i.e., for example, root protein levels, crown bud development, root mass, and others) that seriously confound the results and make it impossible for the investigator to ascertain cause–effect relationships.

To avoid the beforementioned confounding associated with differential cutting management, we have used contrasting alfalfa germplasms and genotypes to examine the relationship between root physiology and legume growth and persistence. Using broad-based alfalfa germplasms, we observed no correlation between genetic differences in shoot regrowth rate and root TNC concentrations (Volenec, 1985). We also evaluated contrasting alfalfa genotypes differing up to four-fold in root starch concentrations, only to find that these lines had similar shoot regrowth rates after defoliation (Fankhauser et al., 1989; Habben & Volenec, 1990). Surprisingly, multiple defoliations that eliminated root starch in low-starch lines did not reduce herbage regrowth rates (Boyce & Volenec, 1992a). One constraint in these studies is that the high and low starch genotypes are not isogenic, and differences in root starch are confounded with other traits inherent to these lines. Nevertheless, no clear cause–effect relationship appears to exist between root TNC concentration and variation in shoot regrowth and persistence. As a result we have begun to examine other physiological and biochemical factors controlling growth and persistence of forage legumes. This has led us to our current interest in root N pools.

## EARLY RESEARCH ON ROOT NITROGEN POOLS IN FORAGE LEGUMES

Until recently few studies have critically evaluated the role of root N pools in regrowth and stress tolerance of alfalfa and other forage legumes. Surprisingly,

Graber et al. (1927) reported reductions in root N concentrations, both in spring as growth resumed, and following harvest. These authors concluded that *both* TNC and N reserves in roots were important for regrowth and persistence of forages; however, because more TNC was lost from roots than was N, TNC reserves may have been viewed as being more important than N reserves in forage shoot regrowth. Later work with sweetclover indicated that it accumulated root N in autumn (Smith & Graber, 1948). In spring, root N concentrations declined from 4.1% on 12 April to 1.4% on 30 May. Defoliation the preceding summer reduced accumulation of root N (and TNC); a situation that was correlated with reduced growth the following spring. Other studies revealed that alfalfa, red clover, and sweetclover all accumulate root N in autumn (Bula & Smith, 1954). Root crude protein (N concentration × 6.25) was positively correlated with winter hardiness of these species. Alfalfa and red clover used 50% of their root N during the last week of April and the first week of May, presumably to support new shoot growth. Bula also examined winterhardiness and root characteristics of 11 sweetclover cultivars differing in cold tolerance (Hodgson & Bula, 1956). All cultivars accumulated high concentrations of root TNC in October, whereas those that survived winter averaged 66% greater concentrations of water-soluble protein N than those that were winter killed. In a second study, Bula et al. (1956) examined root TNC and protein concentrations in alfalfa cultivars differing in winter hardiness. Autumn root TNC concentrations were negatively correlated with cold resistance, while water-soluble protein N was higher in roots of winter hardy alfalfa cultivars. Results from these studies suggest that root N, or a particular root N pool, has a significant impact on forage legume persistence and growth.

Wilding et al. (1960) found that soluble non-amino acid N (primarily soluble protein) averaged 2% of alfalfa root dry weight in summer, and this increased to 2.7% of root dry weight by December, but only in roots of winter hardy alfalfa cultivars. Jung and Smith (1961) boiled aqueous extracts of alfalfa roots to precipitate proteins and separate them from other root N compounds. Concentrations of this protein fraction decreased from 3.5 mg $g^{-1}$ fresh weight on 1 March to 0.5 mg $g^{-1}$ fresh weight on 1 May. Roots of red clover contained one-third as much of this protein fraction as did alfalfa roots, and were not as tolerant of freezing. Concentrations of carbohydrate fractions (sugars, starch, and TNC) were not associated with species differences in freezing tolerance in this study.

Early attempts to characterize alfalfa root proteins using electrophoresis were met with limited success. A group of proteins mobile under native electrophoretic conditions accumulated in alfalfa roots during hardening, however, protein composition of winter hardy and nonwinter hardy lines were similar (Coleman et al., 1966). Gerloff et al. (1967) detected changes in root protein composition that were associated with alfalfa hardening, but contended that these changes were too small to account for the large differences in hardening that occurred. Faw and Jung (1972) also observed slight changes in protein composition of alfalfa roots during hardening, but they concluded that such alterations in protein composition might be sufficient to enhance cold tolerance. Krasnuk et al. (1978) later showed that, in addition to a general accumulation of protein in alfalfa roots during hardening, new isoforms of specific enzymes also were synthesized. Using in vitro assays, Duke and Doehlert (1981) found that activities of most enzymes they assayed were greater in roots of hardy alfalfa varieties during autumn when compared with nonhardy varieties. They speculated that because dinitrogen ($N_2$) fixation of nonhardy varieties was low in autumn, N deficiency could lead to poor acclimation.

## RECENT STUDIES ON ROOT NITROGEN POOLS IN FORAGE LEGUMES

Nitrogen nutrition studies with alfalfa and other forage legumes suggested that reserve N pools may exist in roots and/or crowns. Rates of dinitrogen fixation of alfalfa declined within 24 h of defoliation, and remain low for up to 21 d (Vance et al., 1979; Cralle & Heichel, 1981; Vance & Heichel, 1981; Fishbeck & Phillips, 1982; Macdowall, 1983). Early regrowth, comprised largely of N-rich leaves (Etzel et al., 1988), precedes recovery of dinitrogen fixation. Addition of N as $NO_3^-$ immediately after defoliation has no effect on herbage regrowth rate because uptake of inorganic N from the soil also is reduced by defoliation (Vance & Heichel, 1981). These authors suggested that N remobilization from nodules and from storage pools in roots must be sufficient to meet the N needs of regrowing shoots until $N_2$ fixation resumes. Marriott and Haystead (1992) reached a similar conclusion in their work with N nutrition of white clover. Results of N turnover studies using $^{15}N$ fertilization of subterranean clover (*Trifolium subterraneum* L.) also support this suggestion. Phillips et al. (1983) found that essentially the entire N pool present in crowns and roots at the first forage harvest was replaced by new N by the fourth harvest. These authors suggested that root protein degradation provided N required for leaf growth after defoliation because $N_2$ fixation and soil N uptake were impaired. Culvenor and Simpson (1991) reported that most of the N found in new leaves of regrowing subterranean clover was mobilized from roots and crowns. Soluble proteins in roots supplied most of the mobilized N.

Root N reserves began to interest us when we realized that, like root TNC reserves, alfalfa root protein concentrations varied seasonally (Volenec et al., 1991). Recent findings (Li et al, 1996) indicate that N concentrations increase markedly in autumn in roots of all legume species examined to date, with increases in birdsfoot trefoil roots being the greatest in relative terms (Fig. 6–1A). Substantial reductions in root N occurred for all species as shoot growth resumed in late March. Defoliation on 2 June had little impact on root N concentrations.

When root N was partitioned into its various component pools (Fig. 6–2), we were surprised to learn that the largest single N pool (45%) was not soluble in buffer, with lesser amounts (30%) of low molecular weight N (soluble in trichloroacetic acid, primarily amino acids, inorganic N, and others) and buffer-soluble protein N (including VSPs, 25%; Fig. 6–2; Barber et al., 1996). Expressed as a percentage of total root N, the buffer-insoluble N pool increased after defoliation, suggesting that it is not mobilized to shoots as readily as other N pools after harvest (Barber et al., 1996). In contrast, large changes in soluble protein and low molecular weight N pools of roots occur seasonally and after defoliation (Fig. 6–1B and 6–3). Amino N and soluble protein N accumulate in

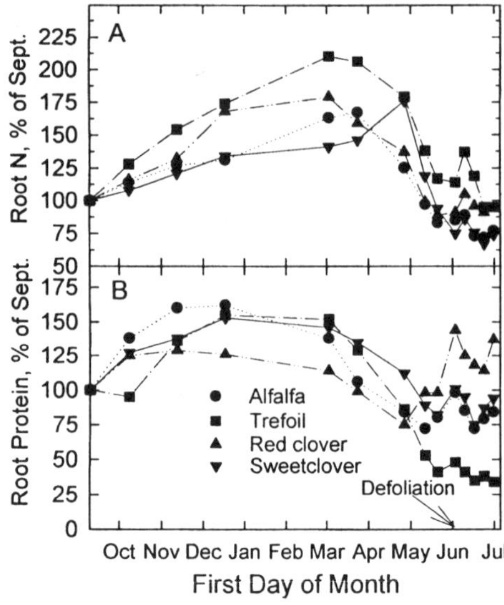


Fig. 6–1. Seasonal changes in root (A) N and (B) root protein concentrations of alfalfa, red clover, sweetclover, and birdsfoot trefoil. Shoot growth resumed in early March, and plants were defoliated 2 June. Data are expressed as a percentage of concentrations present in roots in September (from Li et al., 1996).

autumn as forage legumes harden for winter (Hendershot & Volenec, 1993a; Li et al., 1996). Both of these N pools are rapidly depleted from roots beginning in March as spring shoot growth resumes. Defoliation results in a cyclic pattern of depletion and reaccumulation of root soluble protein and amino N (Fig. 6–1B and

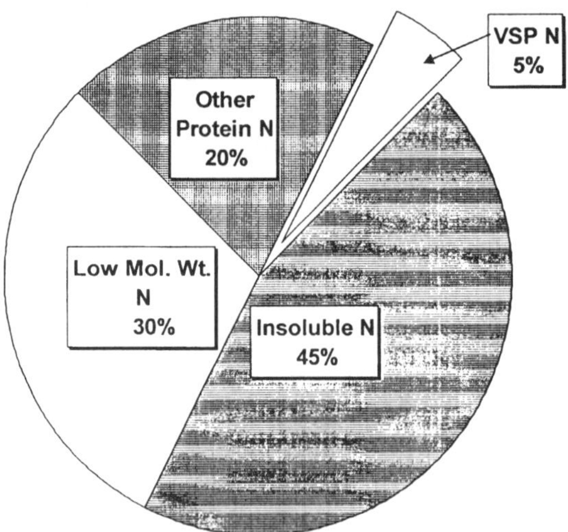


Fig. 6–2. Relative abundance of alfalfa root N pools expressed relative to total root N. Insoluble N was that which is not solubilized in phosphate buffer (pH 7.0); low molecular weight N was soluble in high concentrations of trichloroacetic acid used to precipitate the protein N. The vegetative storage protein (VSP) N was determined using methanol extraction. Details of the procedures used are reported elsewhere (Barber et al., 1996).

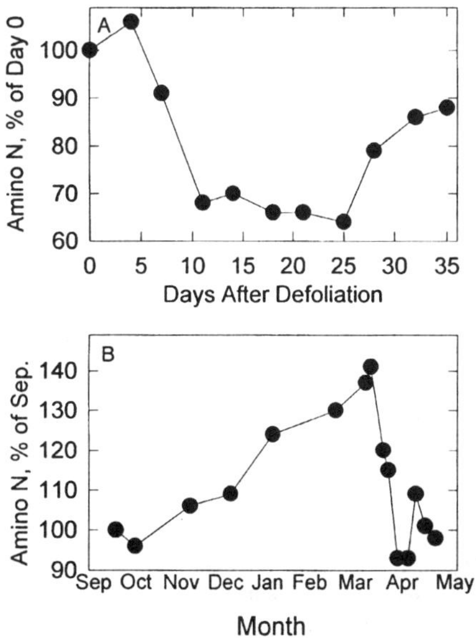


Fig. 6–3. Changes in amino N concentrations of alfalfa roots (A) after defoliation and (B) seasonally. Data are expressed as a percentage of concentrations present in roots at (A) defoliation or in (B) September (from Hendershot & Volenec, 1993a,b).

6–3A), a trend that mimics changes in root TNC concentrations after harvest (Hendershot & Volenec, 1993b; Li et al., 1996). Aspartate and asparagine are very abundant in alfalfa roots comprising >50% of the total amino acid pool (Hendershot & Volenec, 1993b). These amino acids also undergo the largest reduction in concentration after defoliation.

Labeling studies have shown that N lost from alfalfa roots after defoliation is transferred to regrowing shoots (Kim et al., 1991; Ourry et al., 1994; Avice et al., 1996b; Barber et al., 1996). Nearly 80% of the N found in shoots during the initial week of regrowth is derived from N reserves in roots and crowns (Fig. 6–4). This proportion declines later as nodule function is restored and $N_2$ fixation

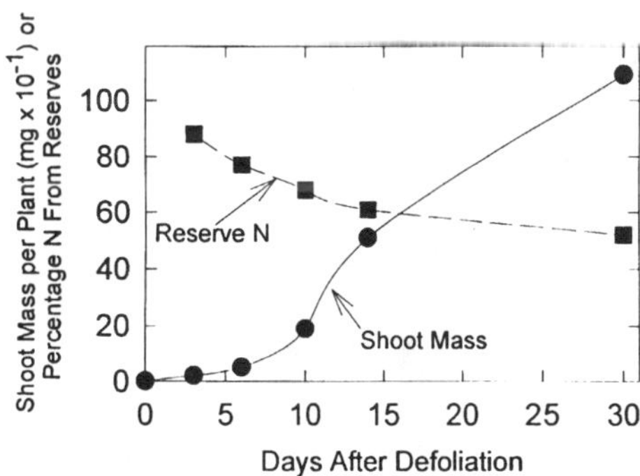


Fig. 6–4. Changes in shoot mass (× $10^{-1}$) and percentage of shoot N derived from root and crown reserves after defoliation of alfalfa (from Avice et al., 1996b).

begins to supply N to meet plant N needs. Nevertheless, after 30 d of growth nearly 50% of the N found in herbage was derived from reserve N pools. Although substantial amounts of labeled C are lost from roots and crowns after harvest, little of this labeled C is transferred from roots and crowns to regrowing shoots (Avice et al., 1996b; Fig. 6–5). The difference between label found in shoot tissues and that lost by roots and crowns is presumably lost via dark respiration, and accounted for nearly 90% of the label disappearing from roots and crowns. In addition, up to 58% of the labeled C compounds found in regrowing shoots was derived from C initially present in N-containing compounds from roots and crowns. This suggests that root N reserves supply N and significant amounts of C to regrowing alfalfa shoots after defoliation. In contrast, C reserves in roots and crowns appear to primarily support dark respiration.

Roots of alfalfa appear to be the primary source of N for regrowing shoots because they undergo the greatest decline in N and $^{15}N$ abundance during shoot regrowth after defoliation (Fig. 6–6). When averaged across fall dormant and nondormant cultivars, roots contained 80% of the N and 71% of the $^{15}N$ at defoliation. After 10 d of shoot regrowth the root N content declined to 58% of the total found in the plant, while the $^{15}N$ content of roots decline to 41% of the plant total. These losses in root $^{15}N$ account for nearly all the $^{15}N$ that appeared in shoots during regrowth. When averaged across dormant and nondormant cultivars, crowns contained 20% of the N and 29% of the $^{15}N$ found in plants at defoliation. After 10 d of shoot growth the level of N in crowns did not change significantly, while the $^{15}N$ level declined slightly to 24% of the total $^{15}N$ found in the plant. Because of the apparent importance of root N pools in alfalfa shoot regrowth, we have focused our efforts on characterizing root proteins, and their role in providing N to regrowing shoots.

Our initial analysis revealed that four polypeptides were very abundant in crude extracts of alfalfa roots (Volenec at al., 1991; Hendershot & Volenec, 1993a,b). Because of their high abundance, and the cyclic pattern of depletion–reaccumulation of these polypeptides after defoliation and in spring when shoot growth resumed, we considered these polypeptides to be putative

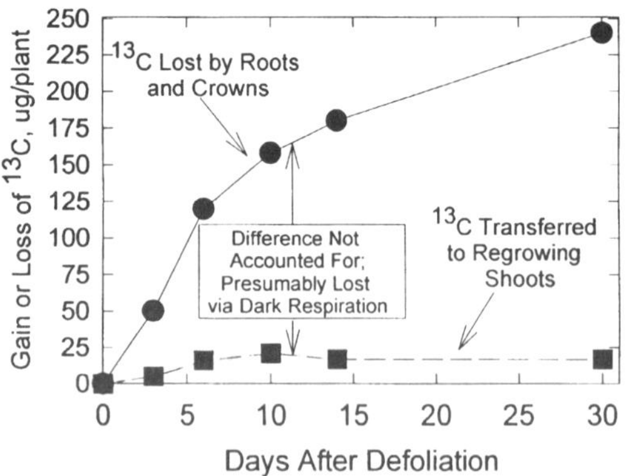


Fig. 6–5. Changes in $^{13}C$ content of roots, crowns, and regrowing shoots of alfalfa after defoliation (from Avice et al., 1996b).

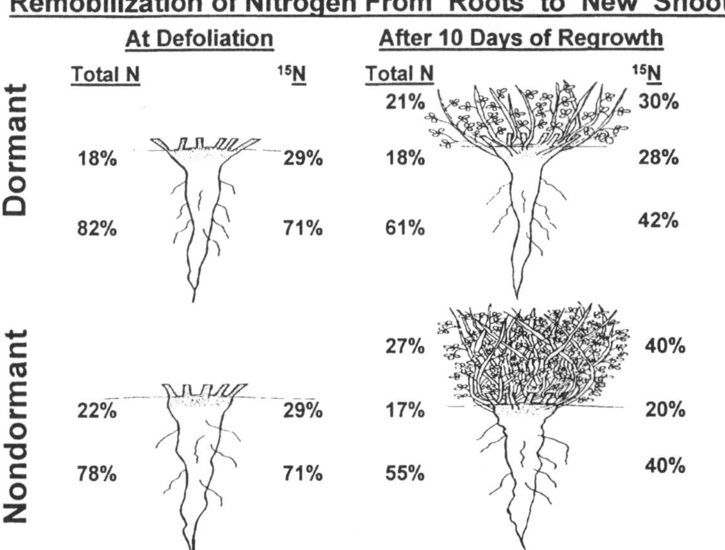


Fig. 6–6. Distribution of N and $^{15}N$ at defoliation and after 10 d of shoot regrowth in fall dormant and nondormant alfalfa (from Barber et al., 1996).

VSPs. Subsequent work indicated that the largest (57.5 kD) of these was β-amylase, which we isolated and characterized (Boyce & Volenec, 1992b). Purification of the other three polypeptides with molecular masses of 15, 19, and 32 kD has recently been reported (Cunningham & Volenec, 1996). We raised antibodies to each putative VSP to serve as molecular probes to specifically ascertain their relative abundance. Efforts to chemically isolate a VSP-enriched fraction from roots

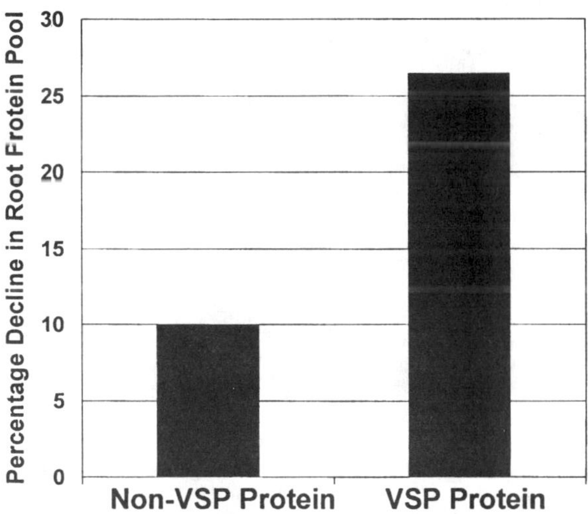


Fig. 6–7. Relative decline in vegetative storage protein (VSP) N and non-VSP N in alfalfa roots during 10 d regrowth of alfalfa (from Barber et al., 1996).

also were successful (Barber et al., 1996). This latter finding allows us to use standard protein analysis techniques to determine VSP concentrations.

Using these tools we showed preferential use of VSPs during shoot regrowth after defoliation (Fig. 6–7). Protein in the VSP-enriched fraction declined about 25% within 10 d of defoliation, while the nonVSP protein pool declined only 10%. Even larger changes in VSP abundance occur between autumn and spring (Fig. 6–8). The VSPs accumulated several-fold between September and December. Little change occurred between December and March, whereas a large decline in all three VSPs occurred between March and early May. By early June, root VSPs reaccumulated to levels that were similar to those present in roots in September. Although some seasonal changes occur in concentrations of other root proteins, most of these changes are minor when compared to changes exhibited by these VSPs.

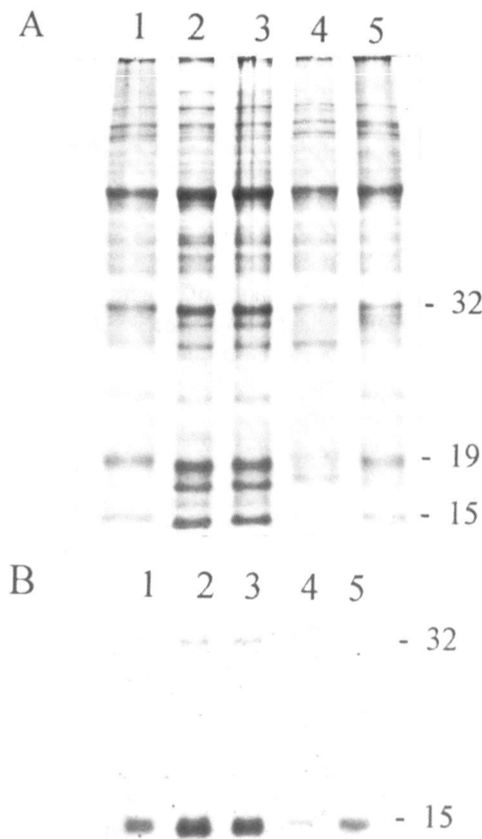


Fig. 6–8. Seasonal changes in abundance of vegetative storage proteins (VSPs) in alfalfa roots. The VSPs have molecular weights of 15, 19, and 32 kD. Panel A: SDS-PAGE analysis of crude extracts of alfalfa roots obtained in September (Lane 1), December (Lane 2), March (Lane 3), May (Lane 4), and June (Lane 5). Panel B: Immunoblot analysis of VSPs using antibody raised to the 15 kD VSP (this antibody also cross-reacts with the 32 kD VSP; from Cunningham & Volenec, 1996).

Although root protein accumulates in other forage legume species during winter hardening (Fig. 6–1b), roots of red clover, birdsfoot trefoil, and sweetclover do not contain polypeptides that function like VSPs. Qualitative analysis of root proteins of these forage legumes revealed no unusually abundant polypeptides in roots of red clover and sweetclover (Fig. 6–9; Cunningham & Volenec, 1996; Li et al., 1996). The very abundant polypeptide (28 kD) observed in trefoil roots is a candidate VSP, however, its concentration does not change in a manner consistent with it being used as an N source when shoot growth resumes in spring or after defoliation (Li et al., 1996). The nature of this protein in trefoil roots remains to be determined. Immunoblot analysis confirmed that alfalfa root VSPs are not found in roots of other forage legume species, but that roots of all perennial members of the *Medicago* genus we have examined to date contain an abundance of these VSPs (Cunningham & Volenec, 1996).

The alfalfa root VSPs are expressed in a tissue-specific manner. Immunoblot analysis of proteins extracted from seeds, stems, and leaves of alfalfa revealed that the root VSPs were not present in these tissues (Fig. 6–10; Cunningham & Volenec, 1996), nor are they present in nodules or small secondary roots of alfalfa (data not shown). The positive immunoblot signal in crown extracts results from the contamination of a small amount of taproot tissue

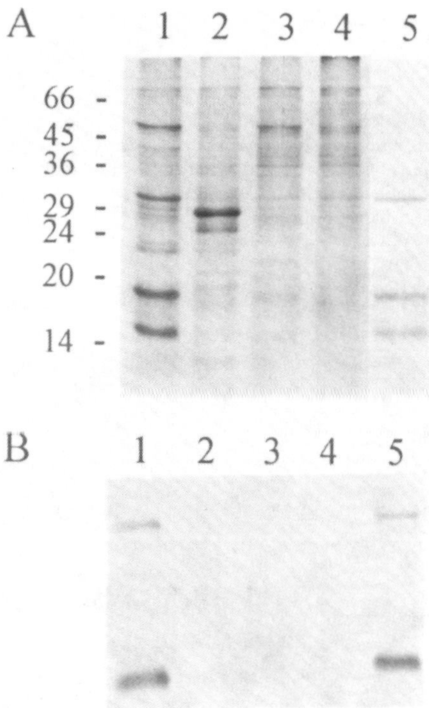


Fig. 6–9. Panel A SDS-PAGE analysis and Panel B immunoblot analysis of crude extracts of proteins from roots of alfalfa (Lane 1), birdsfoot trefoil (Lane 2), red clover (Lane 3), and sweetclover (Lane 4). Purified alfalfa root vegetative storage proteins with molecular wts of 15, 19, and 32 kD were analyzed in Lane 5 (from Cunningham & Volenec, 1996).

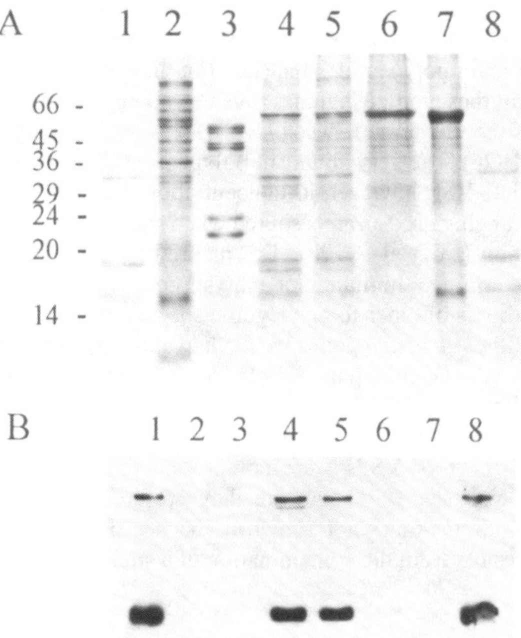


Fig. 6–10. Tissue-specific accumulation of vegetative storage proteins (VSPs) in alfalfa. The VSPs have molecular wts of 15, 19, and 32 kD. Panel A: SDS-PAGE analysis of the following alfalfa tissues: Lane 2, buffer-soluble seed proteins; Lane 3, salt-soluble seed proteins; Lane 4, buffer-soluble root proteins; Lane 5, buffer-soluble crown proteins; Lane 6, buffer-soluble stem proteins; Lane 7, buffer-soluble leaf proteins. Lanes 1 and 8 contain purified root VSPs. Panel B: Immunoblot analysis of VSPs using antibody raised to the 15 kD VSP (this antibody also cross-reacts with the 32 kD VSP; from Cunningham & Volenec, 1996).

Fig. 6–11. The influence of seedling development and cultivar on accumulation of vegetative storage proteins (VSPs) in alfalfa roots. Immunoblot analysis was used to determine relative VSP abundance in roots of seedlings obtained 19, 31, 39, 53, 63, 75, 84, and 123 d after planting. Vernal (fall dormant), Resistar (fall semidormant), and WL 605 (fall nondormant) were used in the analysis (from Kalengamaliro et al., 1997).

attached at the base of the crown, and the high degree of sensitivity of this assay. The mechanisms regulating this tissue-specific expression are unknown, but of great importance. The potential to express genes to very high levels specifically in roots might aid in creating new and innovative approaches to root insect and pathogen resistance, or expression of other genes in roots that could improve legume persistence.

Expression of alfalfa root VSPs also is developmentally regulated (Fig. 6–11). Roots of very young alfalfa seedlings do not contain VSPs. Beginning about 50 d after planting VSPs are detected in seedling roots (Kalengamaliro et al., 1997). The timing of VSP deposition coincides with that of root starch deposition. Onset of deposition differs among cultivars, with earlier VSP deposition occurring in roots of nondormant seedlings (Fig. 6–11, WL 605) that often exhibit faster seedling development when compared with fall dormant alfalfa seedlings (Fig. 6–11, Vernal). Like the tissue-specific expression, we do not yet understand how VSP deposition is regulated developmentally. We are investigating changes in root development that occur during the interval between 30 and 60 d after planting in the hope of identifying cells and intracellular structures involved in VSP accumulation. Results of initial immunochemical localization studies reveal that alfalfa root VSPs are localized in parenchyma cells of wood rays and bark (Avice et al., 1996a). Within cells VSPs are found in vacuoles and on the periphery of starch granules. Therefore, VSP synthesis is closely regulated at the organ, cellular, and intracellular levels. Determining how VSP synthesis and degradation are regulated will improve our understanding of root physiology and biochemistry, legume N and C metabolism, and should ultimately lead us to new and innovative ways to enhance yield, stress tolerance, and persistence of forage legumes

## REFERENCES

Avice, J.C., A. Ourry, G. Lemaire, and J. Boucaud. 1996b. Nitrogen and carbon flows estimated by $^{15}N$ and $^{13}C$ pulse-chase labeling during regrowth of alfalfa. Plant Physiol. 112:281–290.

Avice, J.C., A. Ourry, J.J. Volenec, G. Lemaire, and J. Boucaud. 1996a. Defoliation-induced changes in abundance and immuno-localization of vegetative storage proteins in taproots of *Medicago sativa*. Plant Physiol. Biochem. 34:561–570.

Barber, L.D., B.C. Joern, J.J. Volenec, and S.M. Cunningham. 1996. Supplemental nitrogen effects on alfalfa regrowth and nitrogen remobilization from roots. Crop Sci. 36:1217–1223.

Beuselinck, P.R., J.H. Bouton, W.O. Lamp, A.G. Matches, M.H. McCaslin, C.J. Nelson, L.H. Rhodes, C.C. Sheaffer, and J.J. Volenec. 1994. Improving legume persistence in forage crop systems. J. Prod. Agric. 7:311–322.

Boyce, P.J., and J.J. Volenec. 1992a. Taproot carbohydrate concentrations and stress tolerance of contrasting alfalfa genotypes. Crop Sci. 32:757–761.

Boyce, P.J., and J.J. Volenec. 1992b. β-Amylase from tap roots of alfalfa. Phytochemistry 31:427–431.

Brink, G.E., and G.C. Marten. 1989. Harvest management of alfalfa: Nutrient yield vs. forage quality, and relationship to persistence. J. Prod. Agric. 2:32–36.

Brown, L.G., C.S. Hoveland, and K.J. Karnok. 1990. Harvest management effects on alfalfa yield and root carbohydrates in three Georgia environments. Agron. J. 82:267–273.

Brown, R.H., R.B. Pearce, D.D. Wolf, and R.E. Blaser. 1972. Energy accumulation and utilization. p. 143–184. *In* Alfalfa science and technology. Agron. Monogr. 15. ASA, CSSA, and SSSA, Madison, WI.

Bula, R.J., and D. Smith. 1954. Cold resistance and chemical composition in overwintering alfalfa, red clover and sweetclover. Agron. J. 46:397–401.

Bula, R.J., D. Smith, and H.J. Hodgson. 1956. Cold resistance in alfalfa at two diverse latitudes. Agron. J. 48:153–156.

Coleman, E.A., R.J. Bula, and R.L. Davis. 1966. Electrophoretic and immunological comparisons of soluble root proteins of *Medicago sativa* L. genotypes in the cold hardened and non-hardened condition. Plant Physiol. 41:1681–1685.

Cralle, H.T., and G.H. Heichel. 1981. Nitrogen fixation and vegetative regrowth of alfalfa and birdsfoot trefoil after successive harvests or floral debudding. Plant Physiol. 67:898–905.

Culvenor, R.A., and R.J. Simpson. 1991. Mobilization of nitrogen in swards of *Trifolium subterraneum* L. during regrowth after defoliation. New Phytol. 117:81–90.

Cunningham, S.M., and J.J. Volenec. 1996. Purification and characterization of alfalfa taproot vegetative storage proteins. J. Plant Physiol. 147:625–632.

Duke, S.H., and D.C. Doehlert. 1981. Root respiration, nodulation, and enzyme activities in alfalfa during cold acclimation. Crop Sci. 21:489–495.

Edmisten, K.L., and D.D. Wolf. 1988. Fall harvest management of alfalfa: II. The implications of photosynthesis, respiration, and taproot nonstructural carbohydrate accumulation on fall harvest management. Agron. J. 80:693–698.

Edmisten, K.L., D.D. Wolf, and M. Lentner. 1988. Fall harvest management of alfalfa: I. Date of fall harvest and length of the growth period prior to fall harvest. Agron. J. 80:688–693.

Etzel, M.G., J.J. Volenec, and J.J. Vorst. 1988. Leaf morphology, shoot growth, and gas exchange of multifoliolate alfalfa phenotypes. Crop Sci. 28:263–269.

Fankhauser, J.J., Jr., J.J. Volenec, and G.A. Brown. 1989. Composition and structure of starch from taproots of contrasting genotypes of *Medicago sativa* L. Plant Physiol. 90:1189–1194.

Faw, W.F., and G.A. Jung. 1972. Electrophoretic protein patterns in relation to low temperature tolerance and growth regulation of alfalfa. Cryobiology 9:548–555.

Fishbeck, K.A., and D.A. Phillips. 1982. Host plant and *Rhizobium* effects on acetylene reduction in alfalfa during regrowth. Crop Sci. 22:251–254.

Gerloff, E.D., M.A. Stahmann, and D. Smith. 1967. Soluble proteins in alfalfa roots as related to cold hardiness. Plant Physiol. 42:895–899.

Graber, L.F., N.T. Nelson, W.A. Luekel, and W.B. Albert. 1927. Organic food reserves in relation to growth of alfalfa and other perennial herbaceous plants. Res. Bull. 80. Univ. of Wisconsin, Madison.

Grandfield, C.O. 1943. Food reserves and their translocation to the crown buds as related to cold and drought resistance in alfalfa. J. Agric. Res. 67:33–47.

Habben, J.E., and J.J. Volenec. 1990. Starch grain distribution in taproots of defoliated *Medicago sativa* L. Plant Physiol. 94:1056–1061.

Heichel, G.H., R.H. Delaney, and H.T. Cralle. 1988. Carbon assimilation, partitioning, and utilization. p. 195–228. *In* Alfalfa and alfalfa improvement. Agron. Monogr. 29. ASA, CSSA, and SSSA, Madison, WI.

Hendershot, K.L., and J.J. Volenec. 1993a. Taproot nitrogen accumulation and use in overwintering alfalfa. J. Plant Physiol. 141:68-74.

Hendershot, K.L., and J.J. Volenec. 1993b. Nitrogen pools in taproots of *Medicago sativa* L. after defoliation. J. Plant Physiol. 141:129–135.

Hodgson, H.J., and R.J. Bula. 1956. Hardening behavior of sweetclover (*Melilotus* sp.) varieties in a subarctic environment. Agron. J. 48:157–160.

Jung, G.A., and D. Smith. 1961. Trends in cold resistance and chemical changes over winter in the roots and crowns of alfalfa and medium red clover: I. Changes in certain nitrogen and carbohydrate fractions. Agron. J. 53:359–364.

Kalengamaliro, N.E., J.J. Volenec, S.M. Cunningham, and B.C. Joern. 1997. Seedling development and deposition of starch and storage proteins in alfalfa roots. Crop Sci. 37:1194–1200.

Kim, T.H., A. Ourry, J. Boucaud, and G. Lemaire. 1991. Changes in source–sink relationship for nitrogen regrowth of lucerne (*Medicago sativa* L.) following removal of shoots. Aust. J. Plant Physiol. 18:593–602.

Krasnuk, M., F.H. Witham, and G.A. Jung. 1978. Hydrolytic enzyme differences in cold-tolerant and cold-sensitive alfalfa. Agron. J. 70:597–605.

Kust, C.A., and D. Smith. 1961. The influence of harvest management on the level of carbohydrate reserves, longevity of stands, and yields of hay and protein from Vernal alfalfa. Crop Sci. 1:267–269.

Li, R., J.J. Volenec, B.C. Joern, and S.M. Cunningham. 1996. Seasonal changes in nonstructural carbohydrates, protein, and macronutrients in roots of alfalfa, red clover, sweetclover and birdsfoot trefoil. Crop Sci. 36:617–623.

Macdowall, F.D.H. 1983. Kinetics of first-cutting regrowth of alfalfa plants and nitrogenase activity in a controlled environment with and without added nitrate. Can. J. Bot. 61:2405–2409.

Marriott, C.A., and A. Haystead. 1992. The effect of lenient defoliation on the nitrogen economy of white clover: The contribution of mineral nitrogen to plant nitrogen accumulation during regrowth. Ann. Bot. (London) 69:429–435.

May, L.H. 1960. The utilization of carbohydrate reserves in pasture plants after defoliation. Herb. Abstr. 30:239–245.

Ourry, A., T.H. Kim, and J. Boucaud. 1994. Nitrogen reserve mobilization during regrowth of *Medicago sativa* L. Plant Physiol. 105:831–837.

Pearce, R.B., G. Fissel, and G.E. Carlson. 1969. Carbon uptake and distribution before and after defoliation of alfalfa. Crop Sci. 9:756–759.

Phillips, D.A., D.M. Center, and M.B. Jones. 1983. Nitrogen turnover and assimilation during regrowth in *Trifolium subterraneum* L. and *Bromus mollis* L. Plant Physiol. 71:472–476.

Reynolds, J.H. 1971. Carbohydrate trends in alfalfa (*Medicago sativa* L.) under several forage harvest schedules. Crop Sci. 11:103–106.

Smith, D. 1962. Carbohydrate root reserves in alfalfa, red clover, and birdsfoot trefoil under several management schedules. Crop Sci. 2:75–78.

Smith, D. 1964. Winter injury and the survival of forage plants. Herb. Abstr. 34:203–209.

Smith, D., and L.F. Graber. 1948. The influence of top removal on the root and vegetative development of biennial sweetclover. J. Am. Soc. of Agron. 40:818–831.

Smith, L.H., and G.C. Marten. 1970. Foliar regrowth of alfalfa utilizing $^{14}$C-labeled carbohydrates stored in roots. Crop Sci. 10:146–150.

Vance, C.P., and G.H. Heichel. 1981. Nitrate assimilation during vegetative regrowth of alfalfa. Plant Physiol. 68:1052–1056.

Vance, C.P., G.H. Heichel, D.K. Barnes, J.W. Bryan, and L.E. Johnson. 1979. Nitrogen fixation, nodule development, and vegetative regrowth of alfalfa (*Medicago sativa* L.) following harvest. Plant Physiol. 64:1–8.

Volenec, J.J. 1985. Leaf area expansion and shoot elongation of diverse alfalfa germplasms. Crop Sci. 25:822–827.

Volenec, J.J., P.J. Boyce, and K.L. Hendershot. 1991. Carbohydrate metabolism in taproots of *Medicago sativa* L. during winter adaptation and spring regrowth. Plant Physiol. 96:786–793.

Volenec, J.J., A Ourry, and B.C. Joern. 1996. A role for nitrogen reserves in forage regrowth and stress tolerance. Physiol. Plant. 97:185–193.

Wilding, M.D., M.A. Stahmann, and D. Smith. 1960. Free amino acids in alfalfa as related to cold hardiness. Plant Physiol. 35:726–732.

# 7 Immunological Approaches for Improving Forage Species

**Nicholas S. Hill and Emmett E. Hiatt, III**

*Department of Crop and Soil Sciences*
*University of Georgia*
*Athens, Georgia*

**Tran C. Chanh**

*U.S. Army Medical Research Institute of Infectious Diseases*
*Frederick, Maryland*

## ABSTRACT

Medical research has elucidated a detailed understanding of immunology to where scientific applications are possible in plant sciences. Plant immunology is a relatively recent research discipline, but it has potential to provide powerful tools for diagnostics purposes as well as opportunities for direct improvement of plants via molecular techniques. The purpose of this chapter is to provide a rudimentary understanding of immune function, provide examples of uses of plant immunology in basic and applied research, and describe how plant immunology might be used for forage improvement. Plant immunoassays are based upon polyclonal or monoclonal antibodies, each having defined uses depending upon requirements for specificity of the test being conducted. Monoclonal antibodies are more specific because they are products of individual cell lines with antibody specificity to one epitope of the antigen. Hence, if low molecular weight compounds, such as toxins, are conjugated to immunogenic proteins, it is possible to isolate monoclonal cell lines that produce antibodies to the compound of interest. Monoclonal antibodies can be generated to antigens of specific disease organisms, and the genes responsible for antibody production isolated and inserted into plants to develop novel sources of disease resistance. Similar technology may be useful for providing oral immunizations for livestock grazing forages, thus creating value added forages.

Immunological procedures for improving forage species can be divided into two categories of (i) rapid immunochemical assays for use in plant breeding–genetics–physiology studies and (ii) genetic transformation of plants for antibody induction to reduce plant disease infestation, pharmacological harvesting of the antibodies (e.g., an antibody crop), and possibly to induce antibody production to infectious agents in the grazing animal. Immunochemical assays are powerful


*Molecular and Cellular Technologies for Forage Improvement*. CSSA Special Publication no. 26.

research tools because of their specificity, sensitivity, and versatility for examining or quantifying substances soluble in aqueous solutions. They have distinct advantages over standard analytical techniques or bioassays because they are faster, simpler, less expensive, and environmentally benign (Hill & Agee, 1994; Hill, 1997; Robbins, 1986). Plant pathologists were the first plant scientists to use immunochemical procedures for serologically testing of races of pathogenic viruses in crop species (Dvorak, 1927; Purdy, 1929). Similar procedures still are common practice in modern plant pathology (McDonald & Singh, 1996; Adkins et al., 1996) and have been used to select for virus resistance in crops (Hunger et al., 1991).

The medical sciences expanded the knowledge base of immunology in the 1970s and 1980s (Golub & Green, 1991), and today we are capable of eliciting and purifying antibodies to target compounds not normally considered immunogenic (Robbins, 1986; Weiler, 1986). Applications of immunology to plant science has produced a plethora of literature from which researchers can expand their arsenal of methods to efficiently improve crop species, understand how crops function, and improve the value of crops by engineering them to produce immunogens (Ma & Hein, 1996). Regrettably, there are limited references in American Society of Agronomy (ASA) journals using immunology for crop improvement, suggesting there is either nominal appreciation for immunochemistry and immunology among the Society's clientele, there is little interest in using immunochemical procedures for crop improvement, or other vectors of disseminating crop improvement data are more appropriate than those sponsored by ASA. There is little doubt that publication of crop improvement data is a primary function of the ASA, hence this chapter will focus on the methods and applications of immunology to plant research with the objective of increasing awareness and use of immunology in our discipline.

Immunogens are chemical compounds that are foreign to vertebrate animals and, when used in a vaccine, induce biosynthesis of antibodies (Crumpton, 1974). Not all compounds elicit an immune response when injected, but those that do are usually proteins, and to a lesser extent lipids and complex carbohydrates. For the purpose of this discussion, consider proteins as the model from which a better understanding of immune function can be developed.

## IMMUNE SYSTEM PROCESSING OF IMMUNOGENS: CELL COOPERATION AND ANTIBODY PRODUCTION

When a foreign protein enters the body of a vertebrate animal, three independent cell types recognize the proteins as antigens (antibody promoting; Golub & Green, 1991). The cell types respond through a series of complex and cooperative reactions, culminating in antibody production and secretion to the foreign antigen. There is little need to describe the details of the reactions for the purpose of this discussion, but remedial knowledge of how antibodies are produced will be helpful to develop an understanding of how immunological strategies may be useful for forage improvement.

## Systemic Immunity

Animal cells that produce systemic antibodies circulate in the blood but have their origins in the thymus or bone marrow (Golub & Green, 1991). They are termed T cells and B cells because they are thymus-derived or bone marrow-derived cells accordingly. Immature T cells and B cells are cells that have not yet been exposed to an immunogen and are in a quiescent state until stimulated by an antigenic substance. The quiescent T and B cells work together with a third class of cell, called a macrophage, to initiate the antibody forming process.

Immunogenic proteins are internalized by both macrophages and B cells when present in the bloodstream (Golub & Green, 1991; Harlow & Lane, 1988). Initially, macrophages are more prevalent than B cells and play a vital role in momentum building for completion of the antigenic process. The macrophage cleaves apart the proteins with proteases and presents fragments of the protein back to the cell surface as antigens on specialized structures called major histocompatibility complexes (MHC). Numerous MHC sites occur on each macrophage and each MHC has a unique antigenic site of the immunogen present. Concurrently, B cells perform the same protease cleaving as the macrophage cells, but differ from macrophage cells in that only one fragment is presented to the MHC on the B cell surface. Hence, B cells behave as clonal cell lines that are capable of presenting numerous antigens from a given protein to the cell surface. The B cells also produce weak antibodies, immunoglobulin class IgM, that are low in quantity and have low affinity to the antigen following initial exposure. Since a heterogenous mixture of B cell lines are present in the serum (called polyclonal), they collectively contain a homologous set of antigenic fragments to the immunogen as the antigens on the macrophage MHCs. This results in a polyclonal mixture of IgM antibodies specific to different parts of the immunogenic protein. The B cells grow slowly after initial exposure to the immunogen and serve as a memory cell in the event that a second exposure to the immunogenic protein occurs.

When macrophage cells construct MHCs, quiescent T cells respond to various antigens in at the MHC of macrophage cells and construct T cell receptors (Golub & Green, 1991; Harlow & Lane, 1988). Only one T cell receptor is constructed to an individual MHC by any given individual T cell. Upon division, the T cell serves as a clonal cell line producing only one antigen receptor and, as with B cells, a polyclonal mixture of T cell lines construct a homologous set of antigenic receptors to the antigens present at the MHCs. The T cells are similar to B cells in that they remain in the bloodstream and grow slowly after initial exposure to the immunogenic protein.

When the immunogen enters the body a second time, the macrophage cells process and present antigenic components of the protein in the MHC identically as during initial exposure to the protein (Golub & Green, 1991; Harlow & Lane, 1988); however, the function of B and T memory cells change dramatically. During secondary exposure, T cell receptors attach to the macrophage MHCs that possess antigens specific for the T cell, but rather than producing additional receptors, the T cell produces growth regulating compounds called lymphokines. Rapid proliferation of the T cells with T cell receptors identical to the parent cell

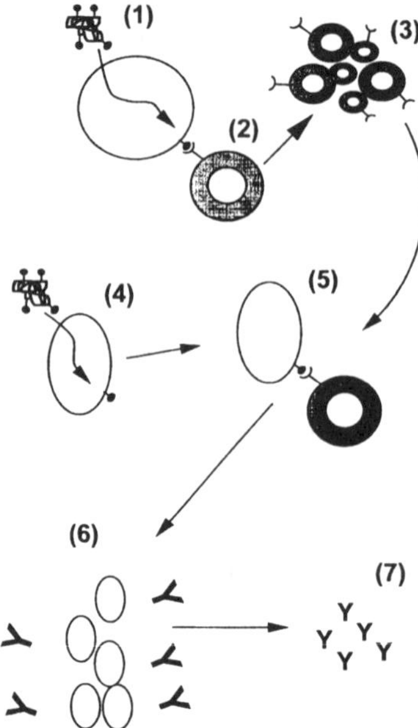


Fig. 7–1. Protein digestion and antigen presentation by macrophages and B cells, and T cell recognition. (1) Protein enters the macrophage, is processed into antigenic fragments, and is presented at the cell surface at the major histocampatibility complex (MHC) (2) The antigen specific receptor of the T cell results in T-cell proliferation (3) Concurrently (4) the B cell processes the protein similarly to the macrophage and builds its own MHC to which the antigen specific receptor of the T-cell can react (5) causing proliferation of B-cells (6) that produces an abundance of antibodies that are serum soluble (7).

results and the abundance of T cells effectively serve as an immunological search party for memory B cells that have MHC antigens compatible with the T cell receptors (Fig. 7–1). When the T cell binds to the memory B cell, a second series

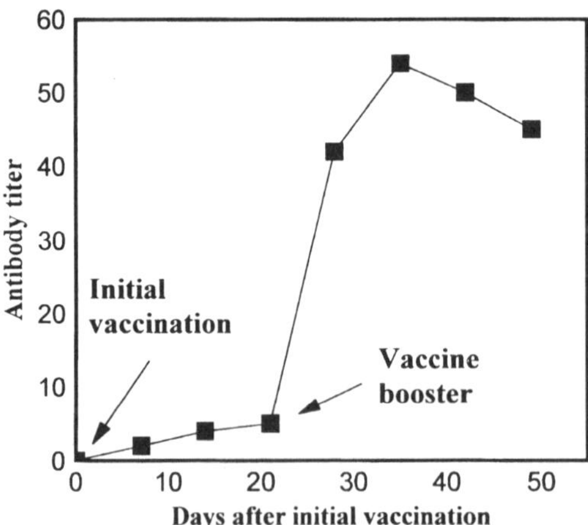


Fig. 7–2. Antibody titer (affinity) in serum in response to initial vaccination and a vaccine booster.

of lymphokines are produced that stimulate rapid proliferation of the B cell. Concomitantly, proliferating B cells change the type of antibodies produced to the immunoglobulin class IgG, which have a higher affinity to the antigen. The B cells produce an abundance of IgG antibodies that are released from the cell, are soluble in the serum, and are capable of binding to the protein containing the antigen for which it is specific. Hence, typical immune responses follow a bimodal antibody affinity to the target immunogen and two vaccinations are necessary to induce antibody production to immunogens (Fig. 7–2).

## Surface Immunity

Mammals possess an extensive array of defense mechanisms within tissues, but it is at surfaces that invading organisms are first encountered and repelled or destroyed. While the skin is an obvious surface, the area of mucosal membranes in the respiratory and intestinal tracts may be 100 times larger than the skin surface and possess an array of immunological protective mechanisms (Tizard, 1996). Some mucosal lymphoid tissues (tonsils, Peyer's patches, intestinal lymphoid nodules, and lung lymphoid nodules) contain T cells, B cells, and macrophages required to initiate an immune response, and operate independently of the systemic immune system. Mucosal lymphoidal tissues constitute 80% of all immunoglobulin producing cells in the body but differ from systemic lymphoidal immunoglobulin producing cells in that they produce primarily a third class of immunoglobulin, IgA. Because of the extent of the mucosal lymphoidal network, IgA antibody production is in excess of IgG antibodies associated with systemic immunity. IgA antibodies are present in saliva, intestinal fluid, nasal and tracheal secretions, tears, milk, cholostrum and urogenital secretions (Bice & Muggenburg, 1996; Bowersock et al., 1992; Russell & Wu, 1991; Santiago et al., 1995; Tizard, 1996; Wu & Russell, 1993). Interestingly, presentation of an antigen at any of the secretory antibody sites results in IgA production at all surface sites.

IgA antibodies are produced in the submucosal cells, are transported across the epithelial cells, and deposited to the cell exterior (Tizard, 1996). IgA can bind to antigens within the submucosa, in the epithelial cells, or at tissue surfaces. A major distinction between IgG and IgA immunoglobulin classes is the association with an independently produced receptor protein, or secretory component (Tizard, 1996). The secretory component of IgA immunoglobulins attaches to the C-terminals of the IgA protein, thereby preventing degradation by proteolytic enzymes associated with intestinal microflora or mucosal tissue. This results in fecal excretion of the antigen–antibody complex, thereby excluding the antigen from circulation. In chickens, rats and rabbits, IgA can be blood-borne and antigen-IgA excreted through the bile into the duodenum.

Mucosal immunity can be elicited by oral vaccination of antigens (Mestecky, 1987; Pierce & Gowans, 1975; Santiago et al., 1995). Single doses of oral vaccines require antigen protection from proteolytic action in the gut. Microspheres (Eldridge et al., 1991), polyacrylamide (O'Hagan, 1989), proteinoids (Santiago, 1995), proteosomes (Orr et al., 1993), and other compounds have been effective antigen delivery systems for IgA stimulation in monogastrics.

Ruminant microflora, however, are capable of fermenting antigen delivery products (Hungate, 1966) and potentially render oral antigen delivery systems useless.

## STRUCTURE OF ANTIBODIES

Antibodies are symmetrical molecules made up of two identical heavy chain subunits (molecular weight = 50 000–75 000 kD) and two identical light chains (molecular weight = 25 000 kD; Goding, 1986). The heavy chains are covalently bound to one another by disulfide bonds and one light chain is bound to each heavy chain by a single disulfide bond (Fig. 7–3). The light chain determines the specificity of the antibody to a target compound, but lysing of cells containing antigens for which the antibody is specific is controlled by the heavy chain. Each light and heavy chain contains a series of homologous units of approximately 110 amino acids that are folded into a domain. The amino acid sequence of domains closest to the N-terminal of both the light and heavy chains vary greatly and is, therefore, called the variable region. The variable regions are the antigen binding site and two variable regions are present on each antibody molecule that are capable of binding to the antigenic agent. The amino acid sequence closest to the C-terminal of the heavy chain is highly conserved and nearly constant. The peptide sequence at the C-terminus contains approximately 15 acidic amino acids followed by approximately 25 nonpolar amino acids. It is presumed that the C-terminus is located in the cytoplasm of the B-cell, the hydrophobic region of the heavy chain spans the lipid bilayer of the B-cell membrane, with the variable regions emanating from the exterior surface of the mem-

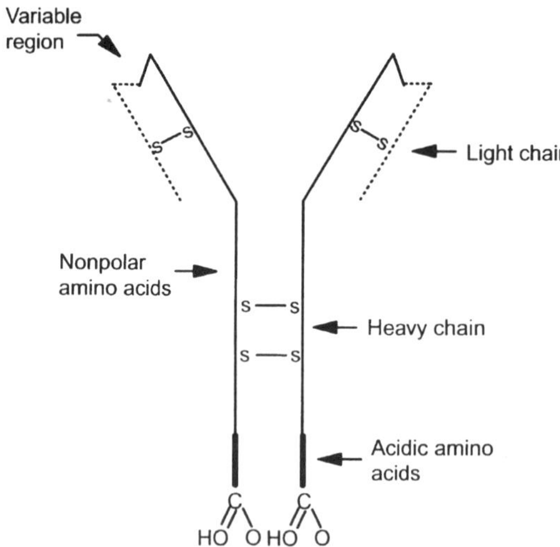


Fig. 7–3. Antibodies are duplicates of a heavy protein chain and a light protein chain. One heavy- and one light-chain are held together by a single disulfide bond between the chains. The N-terminals of the proteins form the variable region that is the binding site to the antigenic agent. The heavy chains are held together by multiple disulfide bonds, thus, each antibody has two variable regions per molecule.

brane. One gene encodes the amino acid sequence of the heavy chain and one gene encodes the amino acid sequence of the light chain.

The antigen-antibody bond is not covalent, but occurs from numerous weak noncovalent interactions from hydrogen bonding, electrostatic forces, van der Waals forces, and hydrophobic effects (Goding, 1986). The bonds are stable across a pH range of 4 to 9 and salt concentration of 0.0 to 1.0 *M*. Therefore, antibody–antigen complexes are stable in isotonic solutions typical of plant and animal fluids.

## ELICITING ANTIBODIES TO LOW-MOLECULAR WEIGHT COMPOUNDS

Molecules with a molecular weight of 1000 or less are incapable of producing natural immune responses (Weiler, 1986). Unfortunately, many compounds of interest to crop scientists (e.g., alkaloids, plant hormones, coumarin, cyanogenic glycosides, estrogenic analogs) do not have the chemical and physical properties necessary to elicit an immune response; however, the immune system can be deceived by covalently conjugating low-molecular-weight compounds to an immunogenic protein. The low-molecular-weight compound thus serves as a hapten unit which, immunologically, will behave as an antigen. Covalent bonds between proteins and low-molecular-weight compounds are conducted using straight chain acids or carbohydrates as chemical bridges; the choice of bridge depending upon the functional groups of the compound to be attached to the protein (Robbins, 1986; Fig. 7–4). Conjugation occurs to free amine radicles of lysine in the peptide chain of the protein. Therefore, the requirements for immunogenicity (large molecular weight, protein in nature) are met by conjugation of low-molecular weight compounds to the immunogenic protein. The protein conjugate is capable of serving as an immunogenic vaccine that is treated as a foreign protein and subjected to cellular internalization, protease activity, and presentation of antigenic units at the MHC of macrophage and B cells. Thus, antibodies will be produced that are specific to the hapten.

Given that many antiquality compounds found in forages have low molecular weights (Fig. 7–5), breeding for reduced quantities of these compounds have generally involved high performance liquid chromotography (HPLC) or colorimetric techniques for quantitation. The speed, low cost, and relative ease with which immunochemical procedures can be used make them advantageous to other methods for forage improvement.

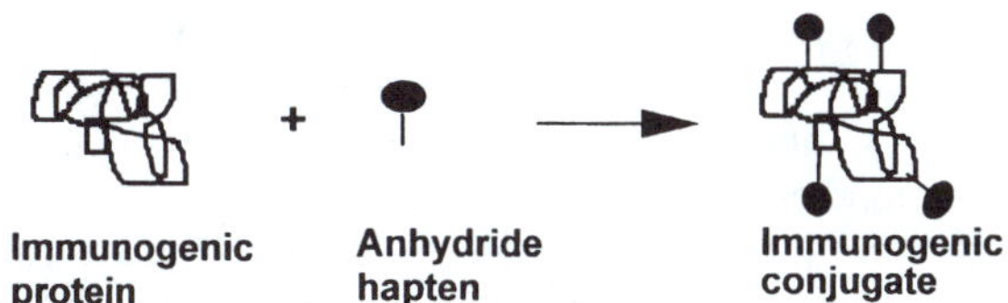


Fig. 7–4. Conjugation of low-molecular-weight haptens to an immunogenic protein can be used to elicit antibodies to the hapten.

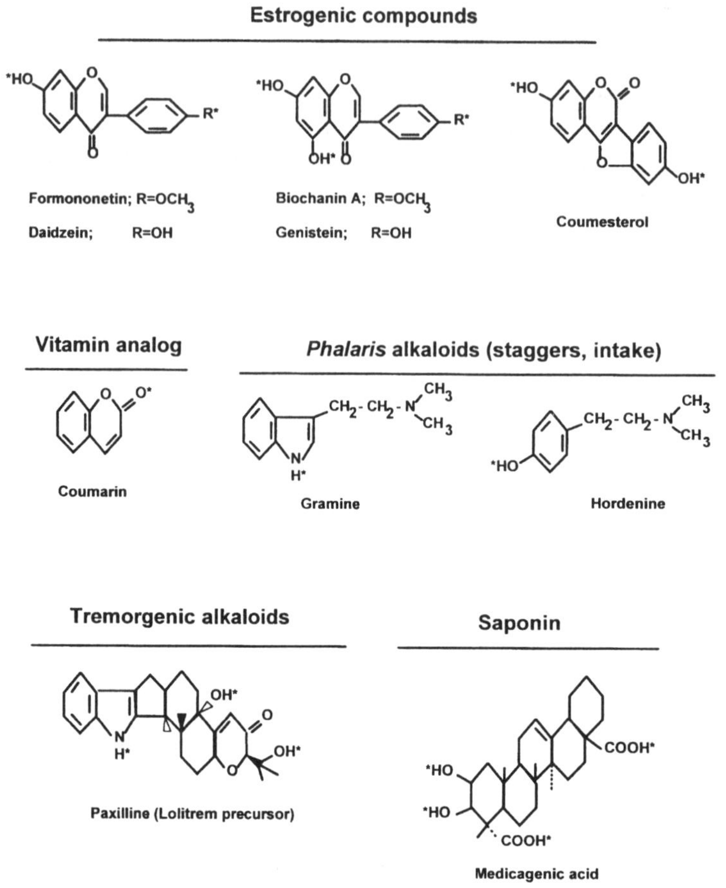

Fig. 7–5. Chemical structures of low-molecular-weight plant derived compounds with functional sites identified (*) to which immunogenic compounds can be conjugated. References for chemical structures are: estrogenic compounds and vitamin analog (Barnes & Gustine, 1973), Phalaris alkaloids (Marten, 1973), saponin (Hanson et al., 1973), and tremorgenic alkaloids (Miles et al., 1995).

## IMMUNOASSAYS FOR PLANT SCIENCES

### Polyclonal vs. Monoclonal Antibodies

Antibodies can be harvested by collecting blood, separating the serum, and the serum containing antibodies diluted and used directly for immunochemical assays. Polyclonal antibodies are generally elicited in laboratory animals such as rabbits or small farm animals (Harlow & Lane, 1988). One admonition when using polyconal antibodies is the cost associated with maintaining the donor animals and the need to revaccinate whenever serum is to be harvested. The immune response varies whenever booster injections are administered, so the serum needs characterization prior to use. The amount of serum that can be obtained from laboratory animals limit use of the antibodies to narrow applications where small

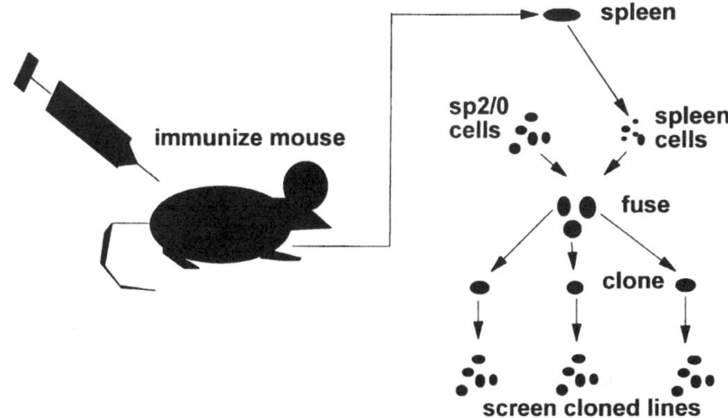


Fig. 7–6. Production of monoclonal antibodies requires harvesting the spleen from the vaccinated mouse, fusion of spleen, and myeloma (sp 2/0) cells, and cloning cell lines. Cell lines are screened for affinity to the immunogen and saved for future use.

quantities of antibodies are needed. Polyclonal antibodies in the serum may nonspecifically bind with other proteins or peptides found in a sample and care must be taken to reduce occurrence of cross reactions. One way to avoid nonspecific cross reactions with nontarget proteins is to cautiously purify the immunogenic protein prior to vaccinating the donor animal; however, the heterogeneous nature of polyclonal antibodies often result in recognition of spurious proteins with antigenic fragments common to the immunogenic protein (Chu, 1991).

A second way to minimize immunological cross reactions is to use antibodies that are produced by a single B cell clone, called a monoclone (Goding, 1986). Monoclonal antibodies are produced by mouse B cells that have been fused with a mouse myeloma cell (Kohler & Milstein, 1975). The resulting hybridoma cell has the perpetual life characteristics of a myeloma cell but the antibody producing capability of the B cell. Because the antibody is from only one B cell line it is specific to only one antigenic region of the immunogenic protein. When generating monoclonal antibodies, B cells are harvested from the spleen, fused with the myeloma cell, and the hybridoma cell lines screened for recognition of the immunogen (Fig. 7–6). Each cell line testing positive to the immunogen can be further characterized for cross reaction with other proteins and selected based upon their specificity to the immunogen. Converse to rigorous purification procedures required prior eliciting polyclonal antibodies, rigorous testing of monoclonal antibodies for cross reactions must be conducted after the cell lines have been selected to avoid false data from the immunoassay (Goding, 1986).

## Types of Immunoassays

Harlow and Lane (1988) provide an excellent review on general uses and attributes of immunoassays and Paraf and Peltry (1991) reviewed immunological techniques used in the food and agricultural industry. The early immunoassays were based upon measuring a precipitated antigen–antibody complex and were simple to conduct (Paraf & Peltry, 1991). Precipitation techniques had limited sensitivity, but current methods have greater sensitivity than immunoprecipitation techniques. For example, the detection limits of indole acetic acid, abscisic acid, cytokinins, and gibberellins ranged in the $10^{-14}$ to $10^{-15}$ molar concentration as compared with gas or liquid chromatography assays with sensitivity ranges in the $10^{-12}$ to $10^{-13}$ molar concentration (Davis et al., 1985). Immunochemical quantitation is 1000 times more sensitive for gibberellin than was gas chromatography equipped with a mass spectrophotometer detection system and is conducted without having to use hazardous solvent systems common to chromatographic procedures.

Regardless of the method used, all modern plant immunoassays have one common feature: the antibody–antigen reaction must be attached to a solid support phase (Harlow & Lane, 1988; Paraf & Peltry, 1991). The immunoassays commonly employed by plant scientists vary in the type of solid support used, and the type of solid support needed generally depends upon the solubility of the antigen to which the antibodies are specific. Solid support systems are available that make immunochemical procedures amenable to antigenic compounds ranging from insoluble to freely soluble in aqueous solutions.

Each major class of immunoassay discussed in this text has numerous variations of methodology. To attempt to represent each variation would be futile, but suffice it to say that variations are designed to increase amplification of the signal by stacking antibodies on top of one another (Harlow & Lane, 1988). This is a process whereby the antigen-specific antibody produced in one animal species has antibodies from another animal species directed to it. Antibodies contain a highly conserved region in the protein that remains constant among antibody types produced within an animal species. The constant region ranges in molecular mass between 50 000 and 75 000, and hence, itself can serve as an immunogen (Goding, 1986). Therefore, the constant region of an IgG antibody from a mouse will elicit mouse IgG-specific antibodies in another animal species. Some immunoassays capitalize on the constant region by using secondary antibodies from another animal that are produced to the constant region of the primary antibody. Secondary antibodies are available commercially and can have fluorescent, chromogenic, or heavy metal entities conjugated to them that permit a visual signal to be produced. Although procedures are available to conjugate recognition reagents to the primary antibody, some antibody activity is lost during the conjugation procedure. Therefore, use of secondary antibodies is preferred. The following immunological methods often vary by the extent of antibody stacking.

## Immunohistology and Immunocytochemistry

Immunohistology and immunocytochemistry are commonly used to detect the presence of an antigenic agent in tissues and cells in situ (Paraf & Peltry, 1991). To do so usually requires that nonantigenic sites of plant tissue be blocked with a novel protein (to which the antibodies will not bind) to prevent nonspecific reactions. Once blocked, insoluble antigens can be detected by bathing cross sections or lateral sections of tissue directly in a solution to permit the primary antibody to anneal to the antigen. Excess antibody is washed from the tissue and a second antibody applied. Secondary antibodies are usually labeled with a fluorescent compound enabling visualization of the antigen in tissue using a fluorescent microscope (Fig. 7–7).

The antibody reaction sequence also can be applied after embedding and sectioning tissues to locate on what organelles an immunogen is present or identify where soluble antigens might be located (Shanklin et al., 1995). The thin sectioned tissues are first blocked with a novel protein, the section exposed to the primary antibody for an extended period (>12 h), the primary antibody washed from the thin section, and a secondary antibody added. Secondary antibodies used for electron microscopic applications usually have a colloidal gold conjugate to visualize antigen. Both of these procedures are used cytologically to localize plant antigenic agents (Freshour et al., 1996; Shanklin et al., 1995; Schuurink et al., 1996), to verify localization of plantibodies (Ma et al., 1995; De Wilde et al., 1996), and examine mycotoxin binding sites in the plant (Jones et al., 1995; Table 7–1).

Miller et al. (1996) used immunocytochemistry to examine progressive degradation of switchgrass (*Panicum virgatum* L.) and big bluestem (*Andropogon gerardii* Vitman) bundle sheath cell (BSC) ribulose carboxylase–oxygenase (RUBPC) degradation the rumen. Their study demonstrated that parenchyma BSC protected RUBPC from degradation during a 24-h period, some of the parenchyma BSC RUBPC exits the rumen without degradation, and protein protected by the BSC escapes and degrades at other sites in the gastrointestinal tract

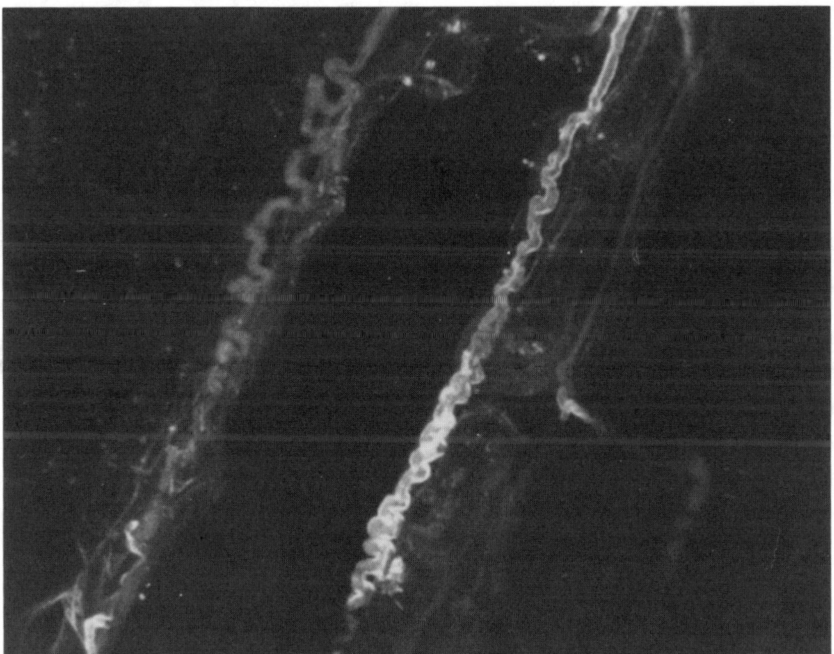

Fig. 7–7. In situ fluorescence of *Neotyphodium coenophialum* produced by an indirect reaction of monoclonal antibodies to surface proteins in the mycelium.

Table 7–1. Recent publications in agricultural and/or biology involving techniques in plant immunology.

| Crop–Plant | Unit of interest | Reference |
|---|---|---|
| Mycotoxin analysis | | |
| *Triticum aestivum* L. | Aflatoxins | DeSaeger & VanPetegham, 1996 |
| *Pinus* sp. | Dothistromin | Jones et al., 1995 |
| *Zea mays* L. | Fumonisin | Shelby et al., 1994; Tejada-Simon et al., 1994; Sutikno et al., 1996; Sydenham et al., 1996 |
| *Hordeum vulgare* L. | Nivalenol | Ikebuchi et al., 1990 |
| Pasture grasses | Sporidesmin | Collins et al., 1995 |
| *Zea mays* L. | Trichothecenes | Park & Chu, 1996 |
| *Brassica* sp. | Zearalenone | Wang & Meng, 1990 |
| Various | Ergot alkaloids | Shelby & Kelly, 1990, 1991, 1992; Hill & Agee, 1994 |
| Pathogen detection | | |
| *Musa acuminata* | Banana bunchy top virus | Geering & Thomas, 1996 |
| *Beta vulgaris* | Beet pseudo-yellow virus | Liu & Duffus, 1990 |
| *Vitus vinifera* | Grapevine fleck virus | Boscia et al., 1995 |
| *Solanum tuberosum* L. | Potato virus Y;A | McDonald & Singh, 1996; Andreeva et al., 1994 |
| *Oryza sativa* L. | *Humicola langinosa* | Dewey et al., 1989, 1990 |
| *Oryza sativa* L. | *Penicillium islandicum* | Dewey et al., 1990 |
| *Glycine max* (L.) Merr. | Soybean mosaic virus | Pacumbara, 1995 |
| *Triticum aestivum* L. | Spindle streak mosaic virus | Carroll et al., 1995 |
| *Triticum aestivum* L. | Barley yellow dwarf virus | D'arcy et al., 1992 |
| Forages | Endophyte | Johnson et al., 1982, 1983, 1985; Musgrave, 1984; Musgrave et al., 1986; Gwinn et al., 1991 |
| Crop physiology | | |
| *Arabidopsis* | Cell-wall polysaccharides | Freshour et al., 1996 |
| *Hordeum vulgare* | Storage protein | Pelger & Bothmer, 1992 |
| Conifers | Seed storage proteins | Misra & Green, 1994 |
| *Zea mays* L. | Cell elongation | Inouhe & Nevins, 1991 |
| Legumes | N-fixing enzymes | Rommeswinkel et al., 1992 |
| *Oryza sativa* L. | Endosperm production | Taira et al., 1991 |
| *Glycine max* (L.) Merr. | Globulin structure | Plumb et al., 1995 |
| *Glycine max* (L.) Merr. | N-fixing enzymes | Triplett & Hao, 1988 |
| *Panicum virgatum* | RuBP carboxylase | Miller et al., 1996 |
| *Triticeae* sp. | Seed storage proteins | Pelger, 1993 |
| *Triticum aestivum* | Seed storage proteins | Curioni et al., 1991 |
| *Triticum aestivum* | Seed storage proteins | Mills et al., 1990 |
| Cell biology | | |
| *Arabidopsis* | Protease ClpP and ClpC | Shanklin et al., 1995 |
| *Hordeum vulgare* | Calmodulin mRNA | Schuurink et al., 1996 |
| *Cucumis anguria* | Cucumber mosaic | Ziegler et al., 1995 |
| *Pyrus communis* | Self-incompatibility genes | Sassa et al., 1991 |
| *Lilium* sp. | Mechanisms of meiosis | Riggs & Hasenkampf, 1991 |
| *Solunum tuberosum* | Hepatitis B (transg. plant) | Domansky et al., 1995 |
| *Glycine max* | Cultivar specific nodulation | Kovacs et al., 1995 |
| Transgenic plants | | |
| *Arabidopsis* | MAK33 (murine antibody against human creatine kinase) | De Wilde et al., 1996 |

(continued on next page)

Table 7–1. Continued.

| Crop–Plant | Unit of interest | Reference |
|---|---|---|
| *Nicotiana tobacum* | Root-knot nematode mAb | Baum et al., 1996 |
| *Nicotiana tobacum* | Phytochrome mAb | Firek et al., 1993 |
| *Nicotiana tobacum* | Murine mAb (Guy's 13) | Ma et al., 1994 |
| *Nicotiana tobacum* | Mucosal immunoglobulin (SigA) | Ma et al., 1995 |
| *Nicotiana tobacum* | Tobacco mosaic virus mAb | Voss et al., 1995 |
| *Nicotiana tobacum* | Artichoke mottled crinkle virus mAb | Tavladoraki et al., 1993 |
| *Nicotiana tobacum* | Botrytis cinerea mAb | van Engelen et al., 1994 |

prior to excretion. They postulated that rumen RUBPC escape could be quantified immunologically at various sites in the digestive tract and used as to examine nutritional significance of the nondegraded protein.

## Immunoblotting

Immunoblotting requires the antigen to be transferred from the plant tissue to a nitrocellulose membrane. The nitrocellulose serves as the solid support for the proteins to which the immunoanalysis can be applied (Fig. 7–8). Plant physiologists are often interested in the location and function of physiologically unique plant proteins. To conduct an immunoblot of the proteins, cellular subunits or organs are isolated and the proteins extracted and separated using polyacrylamide gel electrophoresis. The gels are placed against the nitrocellulose membrane and proteins are electrochemically transferred from the gel to the membrane using a weak electrical current. The membrane is dried, blocked with a nonreactive protein, and bathed in primary antibodies. After washing the membrane, secondary antibodies with a conjugated chromogenic enzyme are permitted to anneal to the antigen-specific antibody. The nitrocellulose membranes are washed and a chromophore solution is added to create a color reaction at the site where the antigen is present. The immunoblot can be compared with a complementary gel in which standard molecular weight proteins were separated to obtain an estimation of the molecular density of the antigen.

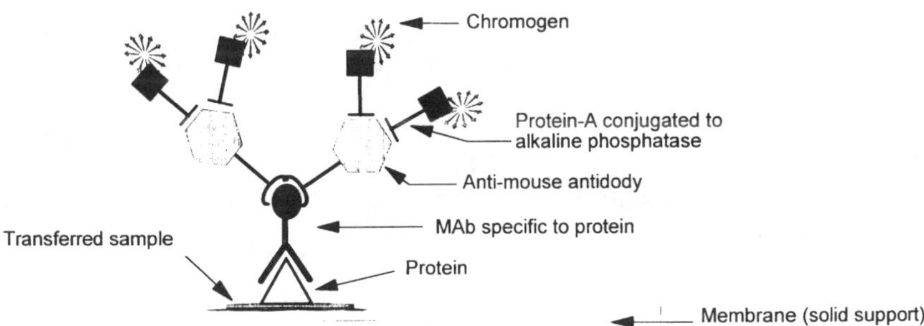


Fig. 7–8. Schematic representation of immunoblotting of antigenic compounds on nitrocellulose membrane.

Polyclonal antibodies are preferred when electrophoretic techniques are used for immunoblotting (Harlow & Lane, 1988). Proteins often are denatured to obtain separation in electrophoretic gels. Using polyclonal antibodies increases the probability of recognition of denaturation-resistant epitopes on the antigen and usually provide a stronger signal than monoclonal antibodies; however, polyclonal antibodies carry an entire complement of antibodies to the antigen and may recognize spurious proteins. Monoclonal antibodies are not widely used because they often will not recognize denatured protein and cross reaction with spurious proteins can be problematic if the antigenic region is present in other proteins; however, several monoclonal antibodies can be pooled to combine the specificity of the monoclonal antibody with the sensitivity of the polyclonal antibody. Immunoblotting is commonly used for detection of storage proteins (Curioni et al., 1991; Pelger & Bothmer, 1992; Misra & Green, 1994), enzymes (Inouhe & Nevins, 1991; Rommeswinkel et al., 1992; Taira et al., 1991), and confirmation of functional plantibodies in transformed plants (De Wilde et al., 1996; Firek et al., 1993; Baum et al., 1996; Table 7–1).

Barber et al. (1996) examined the role of N nutrition on the quantity of 15-, 19-, and 32-kD vegetative storage proteins found in alfalfa roots (*Medicago sativa* L.) and subsequent re- mobilization and utilization following shoot removal. They elicited polyclonal antibodies to each protein fraction and used immunoblotting to verify purity of the protein preparations prior to quantification of each. See Chapter 6 of this special publication for further details on applications of this technology (Volenec, 1998).

Plant pathologists also use immunoblot for qualitative analysis for presence of an infectious agent (Dewey et al., 1990; Table 7–1). This procedure has the option of directly expressing plant sap onto the nitrocellulose membrane for quick screening, or purified proteins can be dotted onto the nitrocellulose membrane if the organism is not sufficiently abundant to be screened in the plant sap. Seed-born pathogens can be analyzed by placing suspect seed directly on the nitrocellulose membrane and letting the antigenic proteins from the pathogens diffuse from the seed to the membrane. The nitrocellulose membrane is dried and an immunoassay similar to immunoblotting is used.

## Enzyme-Linked Immunosorbent Assays

Enzyme-linked immunosorbent assays (ELISA) are used for both qualitative and quantitative measurements of soluble antigenic agents. Three major ELISA methods (direct, indirect, and competitive) are commonly used. All require the antigenic substance to be coated onto a polycarbonate microtiter plate that serves as the solid support. Direct and indirect assays are best for protein antigens and the competitive assays are commonly used for low-molecular weight substances.

## Direct and Indirect Enzyme-Linked Immunosorbent Assays

Direct ELISA is differentiated from indirect ELISA in that the chromogenic enzyme is conjugated directly to the antigen-specific antibody (Harlow & Lane, 1988). Direct ELISA is conducted by coating the bottom of microtiter wells with the antigen, blocking unoccupied sites on the bottom of the well with a novel protein, and incubating the plate with the antigen-specific antibody conjugate. After washing the plate, a chromogenic solution is added to each well and the color reaction measured after a standard time period (Fig. 7–9). Typically horse radish peroxidase or alkaline phosphatase are the chromogenic enzymes conjugated to the antibody and weak peroxide or para-nitrophenylphosphate serve as the respective chromogens. The plates are read with a spectrophotometer and the intensity of the color reaction is stoichiometric with the amount of antigen on the bottom of the plate. Because of the specificity of the antibodies, crude plant extracts can be used to coat the plates, and if desired, a standard curve can be established using varying amounts of purified antigen. Most microplate readers have software capable of calculating a standard curve to convert absorbance values to absolute values. The cost of purification of the antibodies combined with the conjugation process limit the usefulness of direct ELISA. Thus, indirect ELISA assays are more commonly used.

Indirect ELISA is conducted similarly to the direct ELISA, but the chromogenic enzyme is not attached to the antigen-specific antibody. Indirect ELISA uses chromogen conjugated secondary antibodies directed to the antigen-specific antibody (Fig. 7–10). Secondary antibodies are commercially available and inexpensive hence, this is a popular protocol. Antibody stacking provides an ampli-

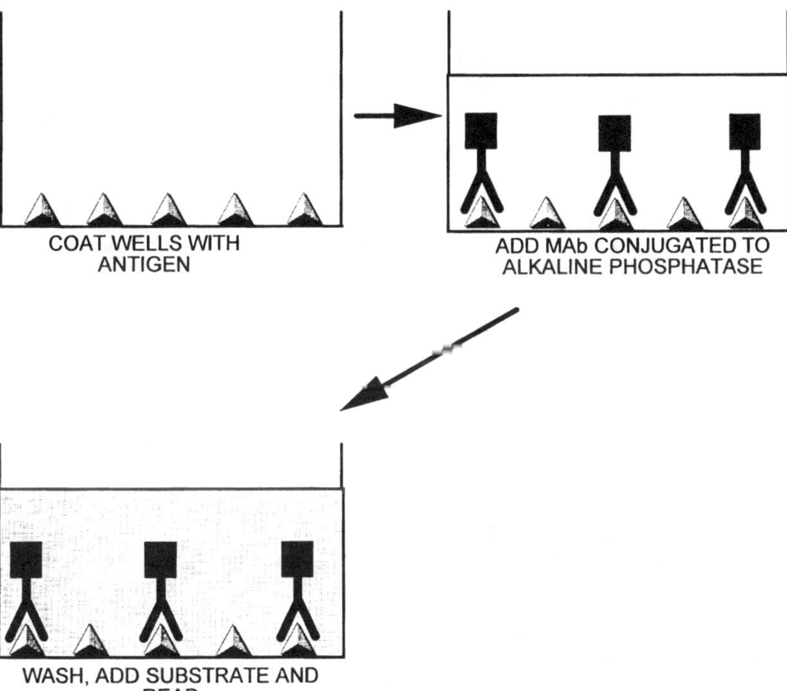


Fig. 7–9. Schematic representation of the sequence of events for conducting direct ELISA to protein antigens in a well from a microtiter plate.

fied signal by the chromogen and provides a more sensitive assay than the direct ELISA. Indirect ELISA is the most commonly used immunoassay among plant scientists. It is used for detecting plant pathogens (Dewey et al., 1990; D'arcy et al., 1992; Liu & Duffus, 1990; Pacumbara, 1995), quantificating pathogenic organisms in the plant (Dewey et al., 1992), confirming of functional antibody in plant tissue (Ma et al., 1994, 1995), quantifying storage proteins (Mills et al., 1990), characterizing enzyme subunits (Pratt et al., 1986; Plumb et al, 1995), and determining taxonomic association among plants (Pelger, 1993; Table 7–1).

## Competitive Enzyme-Linked Immunosorbent Assays

Competitive ELISA is a variation of the indirect ELISA protocol, but is most useful to analyze low-molecular weight, nonimmunogenic compounds. The bottom of the microtiter plate is coated with the low-molecular weight compound that has been conjugated to a protein novel to the donor animal. Plant tissues are typically extracted with a water based solvent or buffer and added to microtiter wells prior to adding the antigen-specific antibody. The antibodies competitively bind to free-floating haptens in the extract and those conjugated to proteins on the plate. If no soluble hapten is present in the buffer, the antibody binds to the hapten conjugate present on the plate. If soluble hapten is present in the buffer, some

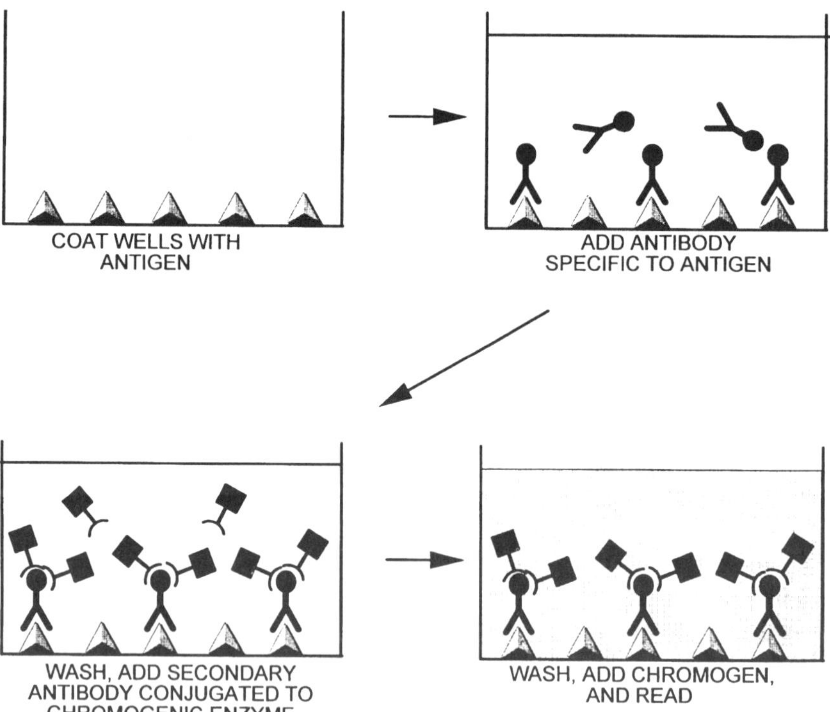


Fig. 7–10. Schematic representation of the sequence of events for conducting indirect ELISA to protein antigens in a well from a microtiter plate.

or all of the antibody binds the hapten in solution preferentially to that which is on the bottom of the plate. Hence, the amount of antibody bound to the conjugate on the bottom of the plate is inversely proportional to that present in the plant (Fig. 7–11) and standard solutions of the soluble compound can be made for quantification. Competitive ELISA is used for mycotoxin detection or quantification (Ikebuchi et al; 1990; Shelby & Kelly, 1990, 1991, 1992; Wang & Meng, 1990; Table 7–1) and plant hormone quantification (Weiler, 1986; Davis et al., 1985).

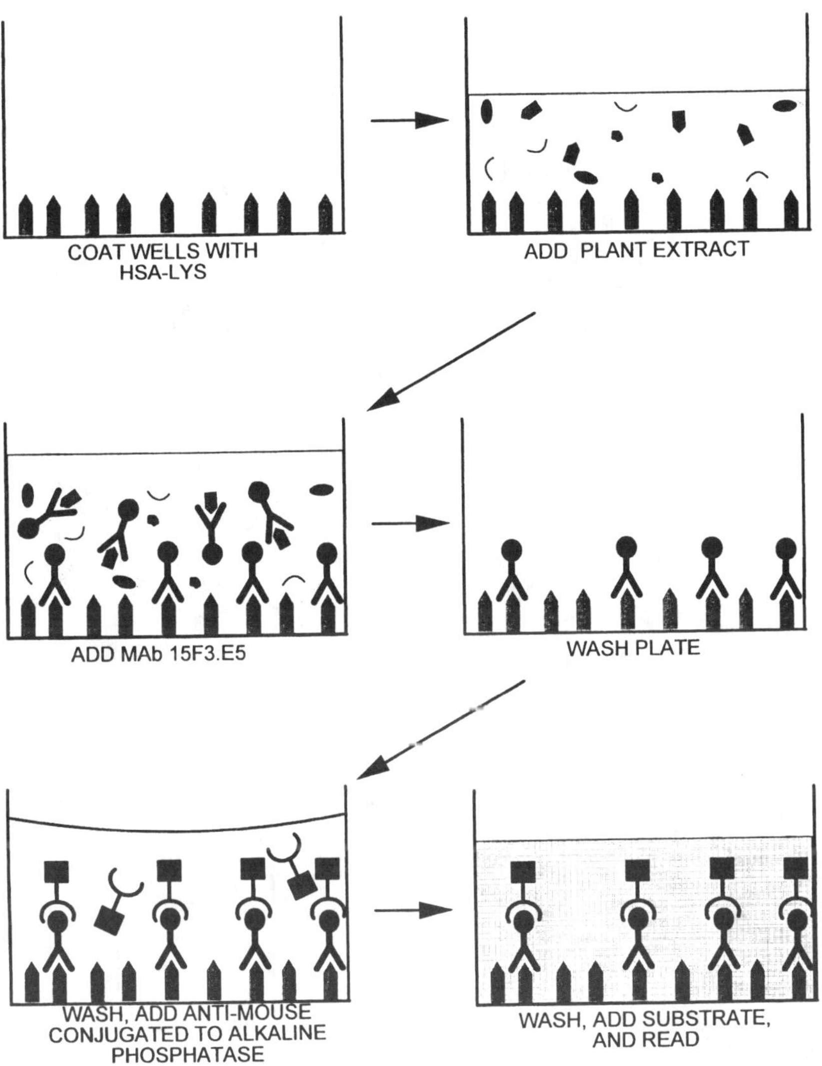


Fig. 7–11. Schematic representation of the sequence of events for conducting competitive ELISA to low-molecular weight compound in a well from a microtiter plate.

Table 7–2. Toxins suspected of causing symptoms of fescue toxicosis in cattle (adapted from Hill et al., 1994).

| Suspect toxin | Function | Mammalian symptoms |
|---|---|---|
| Peramine | Decreased insect herbivory | None known |
| Pyrrolizidine alkaloids | Decreased insect herbivory | Mild vasoconstrictor |
| Paxilline–Tremorgens | Antiungulate herbivory (Perennial ryegrass) | Staggers, tetany |
| Ergot alkaloids | D2 agonist, Antiungulate herbivory | Disorientationreduced conception, vasoconstriction, weight loss |
| Clavine alkaloids | Ergot alkaloid precursor | Similar to ergot |

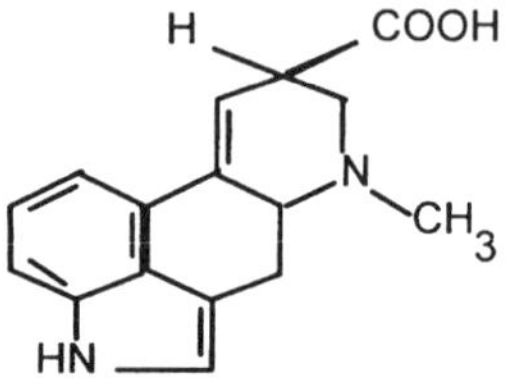

Lysergic acid

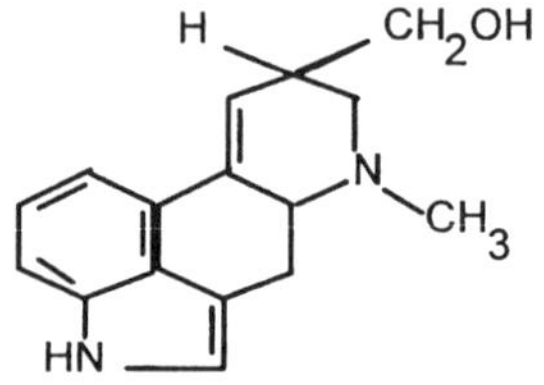

Lysergol

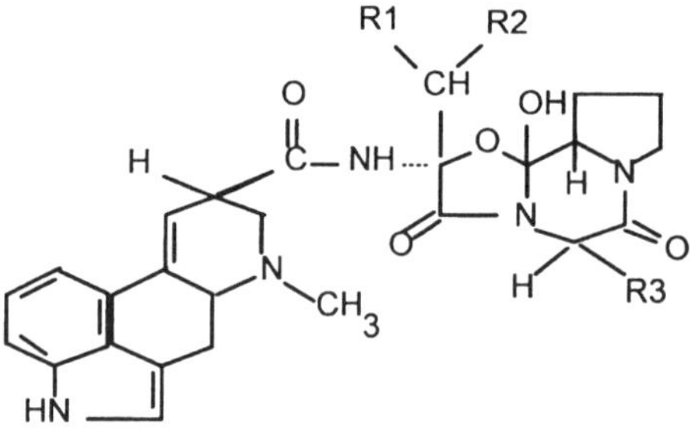

Ergopeptine alkaloids

| R1 | R2 | R3 | | |
|---|---|---|---|---|
| | | $-CH_2C_3H_5$ | $-CH_2CH(CH_3)_2$ | $-CH(CH_3)_2$ |
| H | H | Ergotamine | Ergosine | Ergovaline |
| $CH_3$ | $CH_3$ | Ergocristine | Ergocryptine | Ergocornine |
| H | $CH_3$ | Ergostine | Ergoptine | Ergonine |

Fig. 7–12. Chemical structures of ergot alkaloids found in endophyte-infected tall fescue.

Table 7–3. Molar concentration of lysergic acid derivatives needed to give 50% maximum absorbance when analyzed in a competitive ELISA assay (Hill et al., 1994).

| Lysergic acid derivative | Molar concentration at 50% maximum absorbance |
|---|---|
| Lysergol | $3.93 \times 10^{-15}$ |
| Lysergic acid | $3.73 \times 10^{-11}$ |
| Ergovaline | $1.00 \times 10^{-11}$ |
| Ergonovine | $1.69 \times 10^{-9}$ |
| Ergotamine tartrate | $7.69 \times 10^{-9}$ |
| 2-bromo-a-ergocryptine | $>1.33 \times 10^{-6}$ |
| dihydro-ergocornine | $>1.68 \times 10^{-6}$ |
| dihydro-ergocristine | $>1.78 \times 10^{-6}$ |

## HOW IMMUNOLOGY IS USED IN THE FORAGE IMPROVEMENT PROGRAM AT THE UNIVERSITY OF GEORGIA

A major limitation to livestock production in the southeastern USA is from the effects of fescue toxicosis. Endophyte (*Neotyphodium coenophialum*)-infected tall fescue (*Festuca arundinacea* Schreb.) contains various alkaloids that are candidate toxins presumed to cause the toxicosis syndrome (Table 7–2). Circumstantial evidence suggested that either the ergot or pyrrolizidine alkaloids were the toxic agent (Bush et al., 1979; Lipham et al., 1989; Solomons et al., 1989) but no direct evidence was available to confirm that either was responsible for the condition. The syndrome is difficult to study because many of the clinical symptoms of the syndrome are subjective or semiquantitative, and few animal physiological markers exist from which a definite association between toxicity and animal performance could be made. Prolactin, however, is a circulating hormone in cattle that is depressed whenever animals are exposed to endophyte-infected tall fescue and is, therefore, a good physiological marker to determine if the effects of fescue toxicosis are present in the animal.

Assuming that the ergot alkaloids are causing fescue toxicosis, a monoclonal antibody was generated to the ergoline ring structure common to the ergot alkaloids (Fig. 7–12). By conjugating lysergol to human serum albumin, an ergoline specific immune response was created in a mouse, and a monoclonal cell line identified (15F3.E5) that produced antibodies to the ergoline moiety of the alkaloids (Hill et al., 1994). The antibody did not recognize ergot alkaloids that had hydrogenated or bromated ergoline rings (Table 7–3). Therefore, the antibodies were specific to intact ergoline rings only. Cattle grazing endophyte-infected tall fescue were infused with the monoclonal antibody to sequester circulating ergot alkaloids and serum prolactin measured as an indication of amelioration of the toxicosis syndrome. Prolactin increased within 30 min of antibody infusion and continued to do so for 5 h (Fig. 7–13). Therefore, a direct cause–effect relationships was established between ergot alkaloids and fescue toxicosis using an application of plant immunochemistry.

It is widely recognized that livestock producers are in a biological dilemma when deciding to use endophyte-infected or endophyte-free tall fescue because of superior persistence imparted to the plant by the endophyte (Read & Camp, 1986; Hill et al., 1991; West et al., 1993). Researchers are attempting to develop endo-

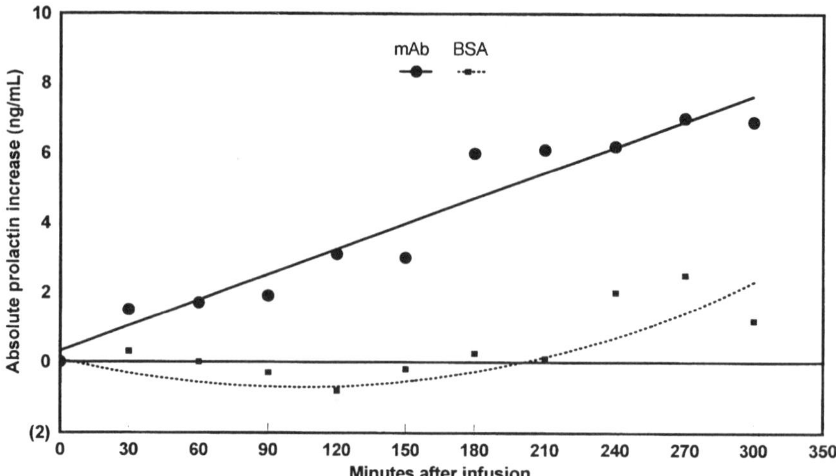


Fig. 7–13. Increase in serum prolactin concentration in response to infusion of a lysergic moiety-specific monoclonal antibody (treated steers) or bovine serum albumin (control steers).

phyte-infected tall fescue that is void of ergot alkaloids. To test a hypothesis that the plant modified to what extent alkaloids could be produced by the endophyte, a reciprocal cross was conducted between low- and high-ergovaline genotypes of tall fescue (Agee & Hill, 1994). They determined that alkaloid concentration was dependent upon the plant genotype even though the alkaloids were produced by the endophyte. From this they postulated that breeding for low-alkaloid endophyte-infected tall fescue may reduce livestock toxicity while maintaining the persistent trait of the plant–endophyte association. Using a competitive ELISA assay, the ergot alkaloid concentration of endophyte-infected tall fescue was subsequently reduced by 86% after two cycles of recurrent selection (Adcock et al., 1997) (Table 7–4). Heritability of the trait was high in both cycles of selection (0.54 in Cycle 1 and 0.90 in Cycle 2). Therefore, immunological protocols are useful for breeding low molecular weight antiquality compounds in forages.

To provide a perspective as to the speed of immunochemistry for plant breeding, our maximum capacity for HPLC screening of plant samples for ergot alkaloids was 80 samples $d^{-1}$ (Hill et al., 1993) but using monoclonal immunochemistry the throughput increased to 1500 samples $d^{-1}$. Furthermore, the amount of tissue needed for HPLC analysis is 1 g but immunochemical procedures required <100 mg of sample (Belesky & Hill, 1997).

Immunochemistry has been used to examine ecological relationships between the tall fescue plant and the endophyte. One concern when breeding for low-alkaloid endophyte-infected tall fescue is that the delicate balance between plant and endophyte may be disrupted. One mechanism by which the plant may modify ergot alkaloid production by the endophyte is through a plant–pathogen interaction in which the plant population is antagonistic to the endophyte. Monoclonal antibodies were generated to a cell wall protein in *Neotyphodium coenophialum* (Hiatt et al., 1997). By examining regenerants of a single genotype

Table 7–4. Mean ergot alkaloid concentrations of the population and families within low and high ergot alkaloid populations of Jesup Improved tall fescue (Adcock et al., 1997).

| | Cycle 1 | | Cycle 2 | |
|---|---|---|---|---|
| Population | $n$† | Alkaloid | $n$ | Alkal oid |
| | | µg kg$^{-1}$ | | µg kg$^{-1}$ |
| Low | 37 | 1231c‡ | 48 | 237c |
| Base | 25 | 1722b | 48 | 1223b |
| High | 44 | 2304a | 48 | 1662a |

† Number of families (12 progeny per family) used to characterize mean ergot alkaloid concentrations within the low and high population; number of individuals used to characterize alkaloid concentration for the base population of Jesup Improved.

‡ Numbers within columns that are followed by different letters were significantly different at the 0.05 level of probability.

of tall fescue into which different endophytes had been inserted, they were able to test the antagonism hypothesis. Endophytic mycelial mass and total ergot alkaloid concentration in tall fescue tissue were quantified using indirect and competitive ELISA. There was no correlation between alkaloid concentration and mycelial mass (Hiatt & Hill, 1997). In fact, plants containing the endophyte with the lowest amount of alkaloid had the greatest amount of endophyte mycelium. In another experiment, 15 different plant genotypes containing a common endophyte had a regression coefficient of 0.17 between leaf alkaloid concentration and mycelial mass, suggesting there was little antagonism to the endophyte by plants low in ergot alkaloid concentration. These experiments provided insight into the possible ecological consequences of breeding for reduced ergot alkaloid concentration in the endophyte-infected population and suggest that continued effort in reducing alkaloid concentration by plant breeding are justified.

Having knowledge of presence or absence of the endophyte in our breeding populations is critical for progress towards obtaining grazing persistent, nontoxic, endophyte-infected tall fescue cultivars. The seed industry is equally concerned about the presence of endophyte in turf cultivars as some are touted as endophyte enhanced. The standard method for screening endophytes in plant tissue involves peeling epidermal and subtending cells from the inner surface of the leaf sheath, staining the cells with a nonspecific mycelial stain, and microscopically examining the tissue for the typical convoluted shape of the endophyte (Clark et al., 1983). The procedure is time consuming, tedious, and not amenable to large scale testing and requires use of seedling plants at that are 6- to 8-wk old to verify the presence of viable endophyte in seed lots. *Neotyphodium*-specific monoclonal antibodies were developed to conduct immunoblot testing for presence of endophyte in seedling tall fescue. In greenhouse-grown plants, endophyte was detected in plant populations 2 wk after emergence and at 3 wk of age the immunoblot approximated microscopic values of 6-wk old seedling plants (E.E. Hiatt, 1998, personal communication; Fig. 7–14). This represents a significant labor saving opportunity for seed testing laboratories because it is adaptable to standard germination procedures, requires less time for the grow-out than does microscopic staining, and sample preparation is faster than tissue staining.

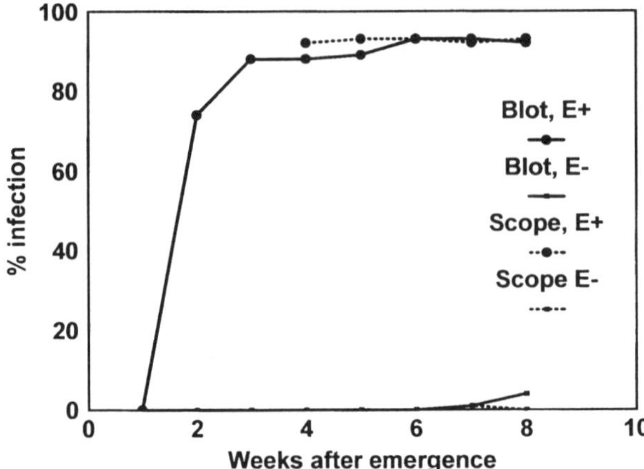


Fig. 7–14. Percentage of infection detected in seedling endophyte-infected (E+) and endophyte-free (E–) plants of Jesup Improved tall fescue using tissue immunoblot of psuedostem cross sections (Blot), or microscopic examination of stained leaf sheaths (Scope).

## IMMUNOLOGY AND PLANT PROTECTION

Techniques for plant transformation using genes encoding murine monoclonal antibodies have been developed and are functional in model plant systems. The coding regions of antibodies are easily detectable in hybridoma cell lines because the primer sequences for the 5' and 3' ends of the heavy and light chains are highly conserved (see Baum et al., 1996 for primer sequence). Therefore, DNA within the two highly conserved regions only produce proteins specific for the antibodies. Once identified, the genes can be ligated into an appropriate vector and plasmids introduced into *Agrobacterium tumifaciens* for plant transformation (Baum et al., 1996; Ma et al., 1994). Individual plants must be transformed each for the heavy and light antibody chains, the transgenic plants cross-pollinated and the $R_1$ plants screened for presence of both chains. Astonishingly, plants are capable of assembling the antibody chains and obtaining a functional antibody with the same affinity as that from the hybridoma cell line (Baum et al., 1996; Ma et al., 1994; Ma et al., 1995; Firek et al., 1993). It is apparent that the antibodies are more stable in plants once they have been assembled as whole units. $R_1$ progeny from the transgenic parents contained as much as 160-fold protein as antibody than either of the $R_0$ parents. Presence of the murine genes and expression of antibodies follows normal Mendelian genetic ratios in transformed populations (Hiatt et al., 1989). The term *plantibody* has adroitly been given to antibodies produced by transgenic plants.

Plantibodies are not expressed in the cytoplasm, rather they accumulate in the apoplasm of the tissue (Baum et al., 1996; Ma et al., 1995). There is a consensus that pantibodies are assembled in the cytoplasm and transported to the exterior of the plant cell but there is no direct evidence to suggest this to be the case (Baum et al., 1996). Presence of plantibodies in the apoplast can have a significant effect on preventing pathological organisms from entering and destroy-

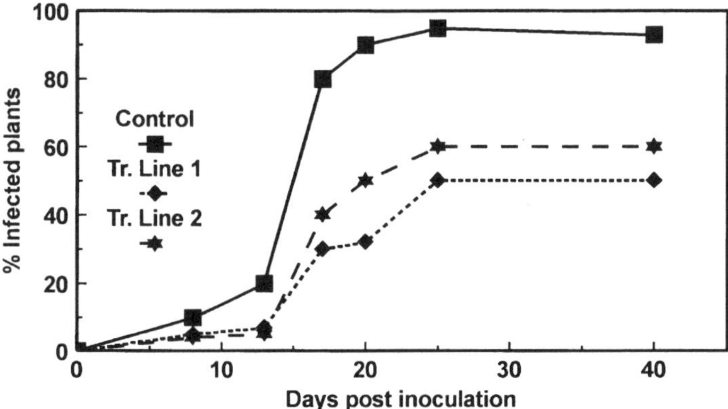


Fig. 7–15. Percentage of infection of transgenic lines of tobacco containing a functional murine antibody to artichoke mosaic crinkle virus (AMCV) following inoculation.

ing plant cells. Transgenic tobacco (*Nicotiana tobacum* L.) plants containing assembled murine antibodies to artichoke mottled crinkle virus (AMCV) were compared with untransformed plants following inoculation with AMCV (Tavladoraki et al., 1993). Nontransgenic plants were >90% infected by AMCV after 15 d but transgenic plants had as low as 50% infection after 40 d (Fig. 7–15). In another experiment, transgenic plants containing assembled murine antibodies to a tobacco mosaic virus (TMV) coat protein were compared with untransformed plants following inoculation with TMV (Voss et al., 1995). Transgenic plants had a 70% reduction in necrotic regions from TMV infection 5 d after inoculation. Furthermore, they found that the extent of protection to TMV was correlated to the amount of anti-TMV murine antibody present in the plants. In this case accumulation of plantibodies in the apoplasm is an advantage because the antigen on the coat protein of the virus is attacked and immobilized prior to entrance of the particle into the plant cell. If the pathogenic antigen is presented to the cytoplasm, thereby escaping association with the plantibody, there is no benefit to the plant. This was demonstrated when functional antibody to root-knot nematode stylet secretions failed to provide protection against infection because stylet secretions are deposited in the cytoplasm (Baum et al., 1996). Strategies for stabilizing functional antibodies in the cytoplasm are being developed for which organisms with direct cytoplasmic infection processes can be treated transgenically (Ma et al., 1995).

## CREATING VALUE ADDED FORAGES THROUGH IMMUNOLOGY

### Anti-Idiotypes

There exists a tremendous opportunity to add value to forage crops by inserting antigens that serve as vaccines to the grazing animal. A monoclonal antibody (MAb1) to an antigen will have two variable regions that are specific to the antigen. If MAb1 is conjugated to an immunogenic protein, it may be possible to revaccinate mice and raise an anti-idiotype monoclonal antibody that is specific to the variable region of MAb1. Antibody–antigen binding requires sufficient conformational complementation between the two molecules for the establishment of binding forces necessary for this interaction to occur. Similarly, conformational complementation allows for the interaction between anti-idiotypic antibodies recognizing the three-dimensional conformation of the variable region of MAb1. As a result of the structural complementation, anti-idiotypic antibodies are recognized serologically as being similar in structure to the nominal antigen, and are capable of antigen mimicry.

Antigen mimicry may provide an opportunity to use anti-idiotype antibodies to substitute for the nominal antigen to induce specific antibody responses. The effectiveness of anti-idiotype vaccines for mimicry of antigens has been documented in systemic antigenic systems (Nissonoff & Lamoyi, 1981) and are capable of inducing antibodies against hepatitis B virus (Kennedy et al., 1984), rabies virus (Reagan et al., 1983), reovirus (Gaulton & Green, 1986), mouse mammary tumor virus (Raychauduri et al., 1986, 1987), and HIV (Attanasio, 1991; Chanh et al., 1989; 1990; Zhou et al., 1990). The advantage of using anti-idiotype vaccines over purified antigens or killed organisms are that (i) hybridoma cell lines are easier and more safe to handle than the live organism, (ii) the surface antigens of these organisms are highly toxic when used in a vaccine and present a health risk to the patient, or (iii) the anti-idiotype is a better immunogen than a nonprotein antigen.

Being a monoclonal antibody, the anti-idiotype gene sequences for the heavy and light chains can be identified and manipulated identically to that of other plantibodies. By inserting the anti-idiotype genes into a forage crop, it may be possible to have the crop manufacture the variable region of the anti-idiotype

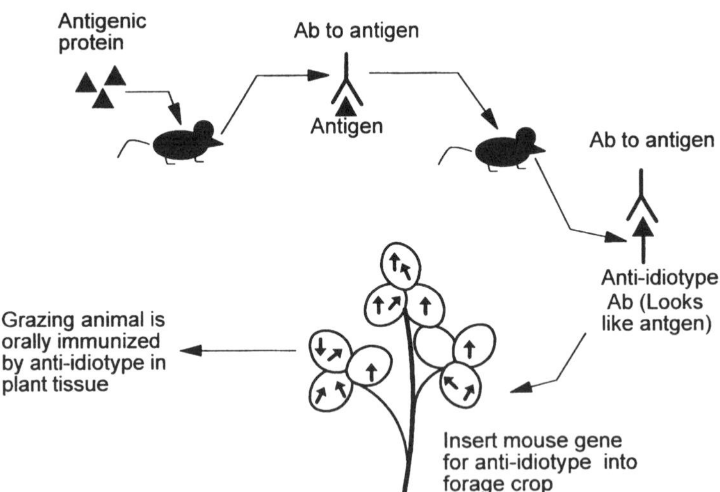


Fig. 7–16. Production of anti-idiotype antibodies to an antigen and insertion of the genes for the variable regions into plants may provide oral vaccines of cattle to infectious disease. This represents a potential mechanism for increasing value of forage crops.

that mimics an antigen. Animals consuming the forage would receive continual oral immunization to the antigen via constant exposure of the anti-idiotype antigen during the eructation and mastication process (Fig. 7–16). Apoplastic presence of the anti-idiotype may be an advantage for leaking of the anti-idiotypic antigens from the plant tissue to the esophagus, a site which induces mucosal immunity. Whether oral dosing of immunogens to the ruminants will result in protective immunity remains to be seen because (i) anti-idiotypes have yet to be demonstrated to provide surface immunity, and (ii) it is unknown whether IgA antibodies are resistant to microbial decomposition in the ruminal–reticular environment.

## Transgenic Plants with Antigens

Transgenic plants have been created that produce surface antigens to hepatitis B (Domansky et al., 1995; Mason et al., 1992), and rabies virus (McGarvey et al., 1995) to elicit immune responses in humans (Mason & Arntzen, 1995). Considering that cows, bulls, and replacement heifers require annual vaccinations for leptospirosis, vibriosis, bovine viral diarrhea (BVD), anaplasmosis, and infectious bovine rhinotracheitis, what might be the prospects for adding value to forages by incorporating surface antigens from livestock diseases through genetic engineering? If genes for surface proteins can be sequenced and the appropriate promoters identified for gene expression, it may be possible to genetically transform forage crops to manufacture the surface proteins to serve as antigenic agents to those organisms. Since these organisms are associated with surface infections, such a strategy appears to be a logical evolution of immunology technology. Gene sequences for surface proteins of BVD are available in the GenBank data base and, hence, may serve as a logical starting point from which plant-derived mucosal immunity can be tested for efficacy in ruminants.

## Antibody Farming

Another pharmaceutical plantibody concept receiving attention is antibody farming (Hiatt & Ma, 1992; Whitelam et al., 1994). Monoclonal antibodies are produced commercially for use as therapeutic agents, cancer treatments, and diagnostic tools for diseases. The estimated value of therapeutic antibodies is in excess of $1 billion dollars. Large-scale production of monoclonal antibodies is complex and expensive for hybridoma cell growth and antibody purification. Incorporation of the antibody genes into forage crops has advantages over laboratory scale production systems in that production is essentially limitless, seeds can be stored easier than hybridoma cell lines or transformed bacterial or yeast cells, perennial forages are buffered against environmental catastrophes, and the product can be efficiently stored as hay. Concentration of plantibody may be a limitation to pharmaceutical antibody farming if simple purification systems are not available for efficient harvest. Antibody farming is yet to reach fruition, but the concept is intriguing.

## REFERENCES

Adcock, R.A., N.S. Hill, J.H. Bouton, H.R. Boerma, and G.O. Ware. 1997. Symbiont regulation and reducing ergot alkaloid concentration by breeding endophyte-infected tall fescue. J. Chem. Ecol. 23:691–704.

Adkins, S., T.J. Choi, B.A. Israel, M.D. Bandla, K.E. Richmond, K.T. Schultz, J.L. Sherwood, and T.L. German. 1996. Baculovirus expression and processing of tomato spotted wilt tospovius glycoproteins. Phytopathology 86:849–855.

Agee, C.S., and N.S. Hill. 1994. Ergovaline variability in *Acremonium*-infected tall fescue due to environment and plant genotype. Crop Sci. 34:221–226.

Andreeva, L., L. Jarvekulg, F. Rabenstein, L. Torrance, B.D. Harrison, and M. Saarma. 1994. Antigenic analysis of potato virus A particles and coat protein. Ann. Appl. Biol. 125:337–348.

Attanasio, R., J.S. Allan, S.A. Anderson, T.C. Chan, and R.C. Kennedy. 1991. Anti-idiotypic antibody response to monoclonal anti-CD-4 preparations in nonhuman primate species. J. Immunol. 146:507–514.

Barber, L.D., B.C. Joern, J.J. Volenec, and S.M. Cunningham. 1996. Effect of supplemental nitrogen on alfalfa regrowth and taproot nitrogen mobilization. Crop Sci. 36:1217–1223.

Barnes, R.F, and D.L. Gustine. 1973. Allelochemistry and forage crops. p. 1–13. *In* A.G. Matches (ed.) Anti-quality components of forages. CSSA Spec. Publ. 4. CSSA, Madison, WI.

Baum, T.J., A. Hiatt, W.A. Parrott, L.H. Pratt, and R.S. Hussey. 1996. Expression in tobacco of a functional monoclonal antibody specific to stylet secretions of the root-knot nematode. Molec. Plant–Microbe Interac. 9:382–387.

Belesky, D.P., and N.S. Hill. 1997. Defoliation and leaf age influence on ergot alkaloids in tall fescue. Ann. Bot. 79:259–264.

Bice, D.E., and B.A. Muggenburg. 1996. Pulmonary immune memory: Localized production of antibody in the lung after antigen challenge. Immunology 80:191–197.

Bioscia, D., V. Elicio, V. Savino, and G.P. 1995. Martelli. Production of monoclonal antibodies to grapevine fleck virus. Plant Pathol. 44:160–163.

Bowersock, T.L., R.D. Walker, M.L. Samuels, and R.N. Moore. 1992. Pulmonary immunity in calves following stimulation of the gut-associated lymphatic tissue by bacterial exotoxin. Can. J. Vet. Res. 56:142–147.

Bush, L.P., J. Boling, and S. Yates. 1979. Animal disorders. p. 247–292. *In* R.C. Buckner and L.P. Bush (ed.) Tall fescue. Agron. Monogr. 20. ASA, CSSA, and SSSA, Madison, WI.

Carroll, J.E. S.M. Gray, and G.C. Bergstrom. 1995. Use of antiserum to a New York isolate of wheat spindle streak mosaic virus to detect related bymoviruses from North America, Europe, and Asia. Plant Dis. 79:346–353.

Chanh, T.C., R.I. Huot, M.R. Schick, and J.F. Hewetson. 1989. Antiidiotypic antibodies against a monoclonal antibody specific for the trichothecene mycotoxin T-2. Toxicol. Appl. Pharmacol. 100:201–207.

Chanh, T.C., G. Rappacciolo, and J.F. Hewetson. 1990. Monoclonal antiidiotype induces protection against the cytotoxicity of the trichothecene mycotoxin T-2. J. Immunol. 144:4721–4728.

Chu, F.S. 1991. Development and use of immunoassays in the detection of ecologically important mycotoxins. p. 87–136 *In* D. Bhatnager et al. (ed.) Handbook of applied mycology. Vol. V. Mycotoxins. Marcel Dekker, New York.

Clark, E.M., J.F. White, and R.M. Patterson. 1983. Improved histochemical techniques for the detection of *Acremonium coenophialum* in tall fescue and methods of in vitro culture of the fungus. J. Microbiol. Methods 1:149–155.

Collins, R.G., L.R. Briggs, and N.R. Towers. 1995. Development and evaluation of an enzyme immunoassay for sporidesmin in pasture. NZ J. Agric. Res. 38:297–302.

Crumpton, M.J. 1974. Protein antigens. p. 1–78 *In* M. Sela (ed.) The antigens. Vol. 2. Academic Press, New York.

Curioni, A., A. Dal Belin Peruffo, G. Pressi, and N.E. Pogna. 1991. Immunological distinction between *x*-type and *y*-type high molecular weight glutenin subunits. Cereal Chem. 68:200–204.

D'arcy, C.J., A.D. Hewings, and C.E. Eastman. 1992. Reliable detection of barley yellow dwarf viruses in field samples by monoclonal antibodies. Plant Dis. 76:273–276.

Davis, G.C., M.B. Hein, B.C. Neely, C.R. Sharp, and M.G. Carnes. 1985. Strategies for the determination of plant hormones. Anal. Chem. 57:638–648.

DeSaeger, S., and C. VanPeteghem. 1996. Dipstick enzyme immunoassay to detect Fusarium T-2 toxin in wheat. Appl. Environ. Microbiol. 62:1880–1884.

Dewey, F.M., M.M. Macdonald, and S.I. Phillips. 1989. Development of monoclonal-antibody-ELISA, -DOT-BLOT and -DIP-STICK immunoassays for *Humicola lanuginosa* in rice. J. Gen. Microbiol. 135:361–374.

Dewey, F.M., M.M. Macdonald, and S.I. Phillips, and R.A. Priestly. 1990. Development of monoclonal-antibody-ELISA and -DIP-STICK immunoassays for *Penicillium islandicum* in rice grains. J. Gen. Microbiol. 136:753–760.

Dewey, F.M., D.R. Twiddy, S.I. Phillips, M.J. Grose, and P.W. Waring. 1992. Development of a quantitative monoclonal antibody-based immunoassay for *Humicola lanuginosa* on rice grains and comparison with conventional assays. Food Agric. Immunol. 4:153–167.

De Wilde, C., M. De Neve, R. De Rycke, A.-M. Bruyns, G. De Jaeger, M. Van Montagu. A. Depicker, and G. Engler. 1996. Intact antigen-binding MAKss antibody and $F_{ab}$ fragment accumulate in intercellular spaces of *Arabidopsis thalina*. Plant Sci. 114:233–241.

Domansky, N., P. Ehsani, A.-H. Salmanian, and T. Medvedeva. 1995. Organ-specific expression of hepatitis B surface antigen in potato. Biotechnol. Lett. 17:863–866.

Dvorak, M. 1927. The effect of mosaic on the globulin of potato. J. Infect. Dis. 41:215–221.

Eldridge, J.H., J.K. Staas, J.A. Meulbroek, J.R. McGhee, T.R. Tice, and R.M. Gilley. 1991. Biodegradable microspheres as a vaccine delivery system. Molec. Immunol. 28:287–294.

Engvall, E., and P. Perlmann. 1971. Enzyme-linked immunosorbent assay (ELISA): Quantitative assay of immunoglobulin G. Immunochemistry 8:871–874.

Firek, S., J. Draper, M.R.L. Owen, A. Gandecha, B. Cockburn, and G. Whitelam. 1993. Secretion of a functional single-chain $F_v$ protein in transgenic tobacco plants and cell suspension cultures. Plant Molec. Biol. 23:861–870.

Freshour, G., R.P. Clay, M.S. Fuller, P. Albersheim, A.G. Darvill, and M.G. Hahn. 1996. Developmental and tissue-specific structural alterations of the cell-wall polysaccharides of *Arabidopsis thaliana* roots. Plant Physiol.110:1413–1429.

Gaulton, G.N, and M.I. Green. 1986. Idiotypic mimicry of biological receptors. Annu. Rev. Immunol. 4:253–286.

Geering, A.D.W., and J.E. Thomas. 1996. A comparison of four serological tests for the detection of banana bunchy top virus in banana. Aust. J. Agric. Res. 47:403–412.

Goding, J.W. 1986. Monoclonal antibodies: Principles and practice. Academic Press, New York.

Golub, E.S., and D.R. Green. 1991. Immunology: A synthesis. 2nd ed. Sinauer Associates, Sunderland, MA.

Gwinn, K.D., M.H. Collins-Shepard, and B.B. Reddick. 1991. Tissue print-immunoblot, an accurate method for the detection of *Acremonium coenphialum* in tall fescue. Phytopathology 81:747–748.

Hanson, C.H., M.W. Pedersen, B. Berrang, M.E. Wall, and K.H. Davis. 1973. The saponins in alfalfa cultivars. p. 33–51. *In* A.G. Matches (ed.). Anti-quality components of forages. CSSA Spec. Publ. 4. CSSA, Madison, WI.

Harlow, E., and D.P. Lane. 1988. Antibodies: A laboratory manual. Cold Spring Harbor Laboratory, Cold Spring Harbor, NY.

Hiatt, A., R. Cafferkey, and K. Bowdish. 1989. Production of antibodies in transgenic plants. Nature (London) 342:76–78.

Hiatt, A., and J.K.-C. Ma. 1992. Monoclonal antibody engineering in plants. FEBS Lett. 307:71–75.

Hiatt, E.E., N.S. Hill, J.H. Bouton, and C.W. Mims. 1997. Monoclonal antibodies for detection of *Neotyphodium coenophialum*. Crop Sci. 37:1265–1269.

Hiatt, E.E., and N.S. Hill. 1997. *Neotyphodium coenophialum* mycelial protein and herbage mass effects on ergot alkaloid concentration in tall fescue. J. Chem. Ecol. 23:2721–236.

Hill, N.S. 1997. Affinity of anti-lysergol and anti-ergonovine monoclonal antibodies to ergot alkaloid derivatives. Crop Sci. 37:535–537.

Hill, N.S., and C.S. Agee. 1994. Detection of ergoline alkaloids in endophyte-infected tall fescue by immunoassay. Crop Sci. 34:530–534.

Hill, N.S., D.P. Belesky, and W.C. Stringer. 1991. Competitiveness of tall fescue as influenced by *Acremonium coenophialum*. Crop Sci. 31:185–190.

Hill, N.S., G.E. Rottinghaus, C.S. Agee, and L.M. Schultz. 1993. Simplified sample preparation for HPLC analysis of ergovaline in tall fescue. Crop Sci. 33:331–333.

Hill, N.S., F.N. Thompson, D.L. Dawe, and F.N. Thompson. 1994. Antibody binding of circulating ergot alkaloids in cattle grazing tall fescue. Am. J. Vet. Res. 55:419–424.

Hungate, R.E. 1966. The rumen and its microbes. Academic Press, New York.

Hunger, R.M., J.L. Sherwood, E.L. Smith, and C.R. Armitage. 1991. Symptomatology and enzyme-linked immunosorbent assay used to facilitate breeding for resistance to wheat soilborne mosaic. Crop Sci. 31:900–905.

Ikebuchi, H., R. Teshima, K. Hirai, M. Sato, M. Ichinoe, and T. Terao. 1990. Production and characterization of monoclonal antibodies to nivalenol tetraacetate and their application to enzyme-linked immunoassay of nivalenol. Biol. Chem. Hoppe-Seyler. 371:31–36.

Inouhe, M. and D.J. Nevins. 1991. Inhibition of auxin-induced cell elongation of maize coleoptiles by antibodies specific for cell wall glucanases. Plant Physiology. 96:426-431.

Johnson, M.C., R.L. Anderson, R.J. Kryscio, and M.R. Siegel. 1983. Sampling procedures for determining endophyte content in tall fescue seed lots by ELISA. Phytopathology 73:1406–1409.

Johnson, M.C., T.P. Pirone, M.R. Siegel, and D.R. Varney. 1982. Detection of *Epichloe typhina* in tall fescue by means of enzyme-linked immunosorbent assay. Phytopathology 72:647–650.

Johnson, M.C., M.R. Siegel, and B.A. Schmidt. 1985. Serological reactivities of endophytic fungi from tall fescue and perennial ryegrass and of *Epichloe typhina*. Plant Dis. 69:200–202.

Jones, W.T., D. Harvey, S.D. Jones, P.W. Sutherland, M.J. Nicol, N. Sergejew, P.M. Debnam, N. Cranshaw, and P.H.S. Reynolds. 1995. Interaction between the phytotoxin dothistromin and *Pinus radiata* embryos. Phytopathology 85:1099–1104.

Kennedy, R.C., K. Alder-Storthz, J.W. Burns, R.D. Henkel, and G.R. Dreesman. 1984. Anti-idiotye modulation of herpes simplex virus infection leading to increase pathogenicity. J. Virol. 50:951–953.

Kohler, G., and C. Milstein. 1975. Continuous cultures of fused cells secreting antibody of predefined specificity. Nature (London) 256:495–497.

Kovacs, L.G., P.A. Balatti, H.B. Krishnan, and S.G. Pueppke. 1995. Transcriptional organization and expression of nolXBTUV, a locus that regulates cultivar-specific nodulation of soybean by *Rhizobium fredii* USDA257. Molec. Microbiol. 17:923–933.

Lipham, L.B., F.N. Thompson, and J.A. Stuedemann. 1989. Effects of metoclopramide on steers grazing endophyte-infected tall fescue. J. Anim. Sci. 67:1090–1097.

Liu, H.-Y., and J.E. Duffus. 1990. Beet pseudo-yellow virus purification and serology. Phytopathology 80:866–869.

Ma, J.K-C., and M.B. Hein. 1996. Antibody production and engineering in plants. Ann. NY Acad. Sci. 792:72–81.

Ma, J.K-C., A. Hiatt, M. Hein, N.D. Vine, F. Wang, P. Stabila, C. Van Dolleweerd, K. Mostov., and T. Lehner. 1995. Generation and assembly of secretory antibodies in plants. Science (Washington, DC) 268:716–719.

Ma, J.K-C., T. Lehner, P. Stabila, C.I. Fux, and A. Hiatt. 1994. Assembly of monoclonal antibodies with IgG1 and IgA heavy chain domains in transgenic tobacco plants. Eur. J. Immunol. 24:131–138.

Marten, G.C. 1973. Alkaloids in reed canarygrass. p. 15–31. *In* A.G. Matches (ed.) Anti-quality components of forages. CSSA Spec. Publ. 4. CSSA, Madison, WI.

Mason, H.S., and C.J. Arntzen. 1995. Transgenic plants as vaccine production systems. Tibtech 13:388–392.

Mason, H.S., P.M.K. Lam, and C.J. Arntzen. 1992. Expression of hepatitis B surface antigen in transgenic plants. Proc. Natl. Acad. Sci. USA 89:11745–11749.

McDonald, J.G., and R.P. Singh. 1996. Host range, symptomology, and serology of isolates of potato virus Y (PVY) that share properties with both the PVYN and PVYo strain groups. Am. Potato J. 73:309–315.

McGarvey, P.B., J. Hammond, M.M. Dienelt, D.C. Hooper, Z.F. Fu, B. Dietzchold, H. Koprowski, and F.H. Michaels. 1995. Expression of the rabies virus glycoprotein in transgenic tomatoes. Bio/Technol 13:1484–1487.

Mestecky, J. 1987. The common mucosal immune system and current strategies for induction of immune response in external secretions. J. Clin. Immunol. 7:265–276.

Miles, C.O., S.C. Munday-Finch, L.P. Meagher, and A.L. Wilkins. 1995. Lolitrem structure: Bioactivity relationships. p. 15–17. *In* L. Garthwaite (ed.) Toxicology and food safety: Research report 1992–1995. AgResearch, Hamilton, New Zealand.

Miller, M.S., L.E. Moser, S.S. Waller, T.J.Klopfenstein, and B.H. Kirch. 1996. Immunofluorescent localization of RuBPCase in degraded C-4 grass tissue. Crop Sci. 36:169–175.

Mills, E.N.C., S.R. Burgess, A.S. Tatham, P.R. Shewry, H.W.S. Chan, and M.R.A. Morgan. 1990. Characterization of a panel of monoclonal anti-gliadin antibodies. J. Cereal Sci. 11:89–101.

Misra, S., and M. Green. 1994. Legumin-like storage polypeptides of conifer seeds and their antigenic cross-reactivity with 11S globulins from angiosperms. J. Exp. Bot. 45:269–274.

Musgrave, D.R. 1984. Detection of an endophytic fungus of *Lolium perenne* using enzyme- linked immunosorbent assay. NZ J. Agric. Res. 27:283–288.

Musgrave, D.R., T.A. Grose, G.C.M. Latch, and M.J. Christensen. 1986. Purification and characterisation of the antigens of endophytic fungi isolated from *Lolium perenne* and *Festuca arundinacea* in New Zealand. NZ J. Agric. Res. 29:121–128.

Nissonoff, A., and E. Lamoyi. 1981. Implications of presence of an internal image of the antigen in anti-idiotypic antibodies: Possible application to vaccine production. Clin. Immunol. Immunopathol. 21:397–406.

O'Hagan, D.T., K. Palin, P. Artursson, S.S. Davis, and I. Sjoholm. 1989. Micorparticles as potentially orally active immunological adjuvants. Vaccine 7:421–424.

Orr, N., G. Robin, D. Cohen, R. Arnon, and G.H. Lowell. 1993. Immunogenicity and efficacy of oral or intranasal *Shigella flexneri* 2a and *Shigella sonnei* proteosome-lipopolysaccharide vaccines in animal models. Infect. Immun. 61:2390–2395.

Pacumbara, R.R. 1995. Seed transmission of soybean mosaic virus in mottled and nonmottled soybean seeds. Plant Dis. 79:193–195.

Paraf, A., and G. Peltry. 1991. Immunoassays in food and agriculture. Kluwer Academic Publ., Boston.

Park, J.J., and F.S. Chu. 1996. Assessment of immunochemical methods for the analysis of trichothecene mycotoxins in naturally occurring moldy corn. J. AOAC Int. 79:465–471.

Pelger, S. 1993. Prolamin variation and evolution in *Triticeae* as recognized by monoclonal antibodies. Genome 36:1042–1048.

Pelger, S., and R. Von Bothmer. 1992. Hordein variation in the genus *Hordeum* as recognized by monoclonal antibodies. Genome. 35:200–207.

Pierce, N.G., and J.L. Gowans. 1975. Cellular kinetics of the intestinal immune response to cholera toxoid in rats. J. Exp. Med. 142:1550.

Plumb, G.W., N. Lambert, E.N.C. Mills, M.J. Tatton, C.C.M. D'Ursel, T. Bogracheva, and M.R.A. Morgan. 1995. Characterisation of monoclonal antibodies against beta-conglycinin from soya bean (*Glycine max*) and their use as probes for thermal denaturation. J. Sci. Food Agric. 67:511–520.

Pratt, L.H., D.W. McCurdy, Y. Shimazaki, and M.M. Cordonnier. 1986. Immunodetection of phytochrome: Immunocytochemistry, immunoblotting, and immunoquantitation. p. 50–74. *In* H.F. Liskens and J.F. Jackson (ed.) Immunology in plant sciences. Springer-Verlag, New York.

Purdy, H.A. 1929. Immunologic reactions with tobacco mosaic virus. J. Exp. Med. 49:919–935.

Rayachaudhuri, S., Y. Saeki, J.J. Chen, H. Iribe, H. Fuji, and H. Kohler. 1987. Tumor specific idiotype vaccines: II. Analysis of tumor related response induced by tumor and by internal image antigens. J. Immunol. 139:271–278.

Rayachaudhuri, S., Y. Saeki, H. Fuji, and H. Kohler. 1986. Tumors pecific idiotype vaccines: I. Generation and characterization of internal image tumor antigen. J. Immunol. 137:1743–1749.

Read, J.C., and B.J. Camp. 1986. The effect of the fungal endophyte *Acremonium coenophialum* in tall fescue on animal performance, toxicity, and stand maintenance. Agron. J. 78:848–850.

Reagan, K.J., W.H. Wunner, T.J. Wiktor, and H. Doprowski. 1983. Antiidiotypic antibodies induce neutralizing antibodies to rabies virus glycoprotein. J. Virol. 48:660–666.

Riggs, C.D., and C.A. Hasenkampf. 1991. Antibodies directed against a meiosis-specific, chromatin-associated protein identify conserved meiotic epitopes. Chromosoma 101:92–98.

Robbins, R.J. 1986. The measurement of low-molecular-weight, non-immunogenic compounds by immunoassay. p. 86–140. *In* H.F. Liskens and J.F. Jackson (ed.) Immunology in plant sciences. Springer-Verlag, New York.

Rommeswinkel, M., B. Karwatski, L. Beerhues, and R. Wiermann. 1992. Immunofluorescence localization of chalcone synthase in roots of *Pisum sativum* L. and *Phaseolus vulgaris* L. and comparable immunochemical analysis of chalcone synthase from pea leaves. Protoplasma 166:115–121.

Russell, M.W., and H.Y. Wu. 1991. Distribution, persistence, and recall of serum and salivary antibody responses to peroral immunization with protein antigen I/II of *Streptococcus mutans* coupled to the cholera toxin B subunit. Infect. Immunol. 59:4061–4070.

Santiago, N., S. Haas, and R.A. Baughman. 1995. Vehicles for oral immunization. p. 413–438. *In* M. Powell and M. Newman (ed.) Vaccine design: The subunit and adjuvant approach. Plenum Press, New York.

Sassa, H., H. Hirano, and H. Ikehashi. 1991. Identification and characterization of stylar glycoproteins associated with self-incompatibility genes of Japanese pear, *Pyrus serotina* Rehd. Molec. Gen. Genet. 241:17–25.

Schuurink, R.C., P.V. Chan, and R.L. Jones. 1996. Modulation of calmodulin mRNA and protein levels in barley aleurone. Plant Physiol. 111:371–380.

Shanklin, J., N.D. Dewitt, and J.M. Flanagna. 1995. The stroma of higher plant plastids contain ClpP and ClpC, functional homologs of *Escherichia coli* ClpP and ClpA: An archetypal two- component ATP-dependent protease. Plant Cell. 7:1713–1722.

Shelby, R.A. and V.C. Kelly. 1990 An immunoassay for ergotamine and related alkaloids. J. Agric. Food Chem. 38:1130-1134.

Shelby, R.A., and V.C. Kelly. 1991. Detection of ergot alkaloids in tall fescue by competitive immunoassay with a monoclonal antibody. Food Agric. Immunol. 3:169–177.

Shelby, R.A., and V.C. Kelly. 1992. Detection of ergot alkaloids from Claviceps species in agricultural products by competitive ELISA using a monoclonal antibody. J. Agric. Food Chem. 40:1090–1092.

Shelby, R.A., G.R. Rottinghaus, and H.C. Minor. 1994. Comparison of thin-layer chromatography and competitive immunoassay methods for detecting fumonisin on maize. J. Agric. Food Chem. 42:2064–2067.

Solomons, R.N., J.W. Oliver, and R.D. Linnabary. 1989. Reactivity of dorsal pedal vein of cattle to selected alkaloids associated with *Acremonium coenophialum* infected fescue grass. Am. J. Vet. Res. 50:235–238.

Sutikno, M.M. Abouzied, J.I. AzconaOlivera, L.P. Hart, and J.J. Pestka. 1996. Detection of fumonisins in *Fusarium* cultures, corn, and corn products by polyclonal antibody-based ELISA: Relation to fumonisin B-1 detection by liquid chromatography. J. Food Protec. 59:645–651.

Sydenham, E.W., G.S. Shephard, P.G. Thiel, C. Bird, and B.M. Miller. 1996. Determination of fumonisins in corn: Evaluation of competitive immunoassay and HPLC techniques. J. Agric. Food Chem. 44:159–164.

Taira, T., M. Uematsu, Y. Nakano, and T. Morikawa. 1991. Molecular identification and comparison of the starch synthase bound to starch granules between endosperm and leaf blades in rice plants. Biochem. Genet. 29:301–311.

Tavladoraki, P., E. Benvenuto, S. Trinca, D. De Martinis, A. Cattaneo, and P. Galeffi. 1993. Transgenic plants expressing a functional single-chain $F_v$ antibody are specifically protected from virus attack. Nature (Washington, DC) 366:469–472.

Tejada-Simon, M.V., L.T. Marovatsanga, and J.J. Pestka. 1994. Comparative detection of fumonisin by HPLC, ELISA, and immunocytochemical localization in *Fusarium* cultures. J. Food Protec. 58:666–672.

Tizard, I.R. 1996. Veterinary immunology: Immunity at body surfaces. W.B. Saunders Co., Philadelphia.

Triplett, E.W., and A.C. Hao. 1988. Molecular and serological comparisons of purified xanthine dehydrogenases from root nodules of three ureide-producing legume species. Physiol. Plant. 74:164–168.

van Engelen, F.A., A. Schouten, J.W. Molthoff, J. Roosien, J. Salinas, W.G. Dirkse, A. Schots, J. Bakker, F.J. Gommers, M.A. Jongsma, D. Bosch, and WJ. Stiekema. 1994. Coordinate expression of antibody subunit genes yields high levels of functional antibodies in roots of transgenic tobacco. Plant Molec. Biol. 26:1701–1710.

Voss, A., M. Niersbach, R. Hain, H.J. Hirsch, Y.C. Liao, F. Kreuzaler, and R. Fischer. 1995. Reduced virus infectivity in *N. tabacum* secreting a TMV-specific full-size antibody. Molec. Breed. 1:39–50.

Wang, H., and F.-J. Meng. 1990. Studies on zearalenone combining substance in the vernalized seeds of *Brassica chinensis* L. Chinese J. Bot. 2:159–164.

Weiler, E.W. 1986. Plant hormone immunoassays base on monoclonal and polyclonal antibodies. p. 1–17. *In* H.F. Liskens and J.F. Jackson (ed.) Immunology in plant sciences. Springer-Verlag, New York.

West, C.P., E. Izekor, K.E. Turner, and A.A. Elmii. 1993. Endophyte effects on growth and persistence of tall fescue along a water-supply gradient. Agron. J. 85:264–270.

Whitelam, C.G., W. Cockburn, and M.R.L. Owen. 1994. Antibody production in transgenic plants. Biochem. Soc. Trans. 22:940–944.

Wu, H.Y., and M.W. Russell. 1993. Induction of mucosal immunity by intranasal application of streptococcal surface protein antigen with the cholera toxin B subunit. Infect. Immunol. 61:314–322.

Zhou, E.M., K.L. Lohman, and R.C. Kennedy. 1990. Administration of non-internal image monoclonal anti-idiotypic antibody induced idiotyope-restricted responses specific for human immunodeficiency virus envelope glycoprotein epitopes. Virology 174:9–17.

Ziegler, A., L. Torrance, S.M. Macintosh, G.H. Cowan, and M.A. Mayo. 1995. Cucumber mosaic cucumovirus antibodies from a synthetic phage display library. Virology. 214:235–238.

# 8 Novel Endophyte Technology: Selection of the Fungus[1]

**C. P. West, M. L. Marlatt, and M. E. McConnell**

*Department of Agronomy*
*University of Arkansas*
*Fayetteville, Arkansas*

**E. L. Piper**

*Department of Animal Science*
*University of Arkansas*
*Fayetteville, Arkansas*

**T. J. Kring**

*Department of Entomology*
*University of Arkansas*
*Fayetteville, Arkansas*

## ABSTRACT

Endophytic fungi of *Neotyphodium* sp. occur naturally in *Festuca* sp. and *Lolium* sp., which are grasses widely managed for forage and turf uses. The endophytes typically produce alkaloids that reduce growth and reproduction in grazing livestock while conferring resistance in the grass hosts to insects, nematodes, and drought stress. Novel endophyte technology involves selecting strains of fungus that do not produce livestock toxins and then introducing them into agronomically desirable host populations to form new compatible associations. The steps involved are to (i) acquire *Festuca* and *Lolium* germplasm from diverse sources and screen for endophyte-containing plants, (ii) select endophyte-infected plants via chomatographic analysis that lack ergot-related alkaloids (livestock toxins), but which contain the insect-deterring alkaloids peramine and lolines, (iii) screen those plants for insect resistance using bioassays, (iv) isolate the most promising endophyte strains from the endemic hosts and transfer to improved cultivars via seedling-stab inoculation, (v) grow out the newly infected plants in the field for seed production and select those strains which show a high compatibility and seed transfer with the cultivar, and (vi) increase seed containing the most desirable strains in large enough amounts to establishing livestock grazing trials. Essentially 100% seed transmission of endophyte is a key criterion for full compatibility and to ensure that novel endophytes have

[1] Manuscript no. 97085. Published with permission of the Director of the Arkansas Agricultural Experiment Station. Supported by Southern Regional IPM Grant no. 91-34103-5832; and Center for Alternative Pest Control, USDA Special Grant no. 95-34195-1330.


*Molecular and Cellular Technologies for Forage Improvement.* CSSA Special Publication no. 26.

commercial viability. A regenerated cultivar containing the new endophyte must be retested for the desirable alkaloid profile. Animal performance and plant persistence trials are critical to demonstrating the success of the novel endophyte in improving a cultivar.

The fungal endophytes of *Neotyphodium* sp. (formerly *Acremonium* sp.) exist in symbioses with *Festuca* sp. and *Lolium* sp. grasses. Associations involving tall fescue (*F. arundinacea* Schreb.) and perennial ryegrass (*L. perenne* L.) have been widely studied owing to livestock disorders caused by toxic alkaloids produced by the endophytes. The toxins most closely associated with fescue toxicosis are the ergot alkaloids (Garner et al., 1993), whereas those causing ryegrass staggers are lolitrem B and its precursor paxilline (Rowan, 1993). Fescue toxicosis and ryegrass staggers cause large economic losses in the USA and New Zealand, respectively (Hoveland, 1993; Prestidge, 1993). Therefore, the top priority in improving tall fescue and perennial ryegrass use by livestock is to arrest production of toxins in the forage or to mitigate their effects in the animal. This can potentially be accomplished at the animal level, the plant level, or the fungal level (Siegel, 1993). Novel endophyte technology involves selecting strains of fungus that do not produce livestock toxins and then introducing them into agronomically desirable host populations to form new compatible associations, or neosymbiota.

An obvious cure for fescue toxicosis and ryegrass staggers would be to sow endophyte-free (E–) seed. Since the endophyte spreads in nature strictly by seed transmission, an E– stand would remain free of endophyte, assuming complete eradication of previously existing, infected (E+) plants and prevention of E+ seed germination or ingress to the field. The endophyte, however, protects its host from some seedling diseases (Gwinn & Gavin, 1992), and herbivory from a wide range of insects, nematodes (West & Gwinn, 1993), and rodents (Coley et al., 1995). Removal of the endophyte *N. lolii* from perennial ryegrass results in rapid stand decimation due to damage by Argentine stem weevil (*Listronotus bonariensis* Kuschel) in New Zealand (Prestidge, 1993). The endophyte *N. coenophialum* promotes tall fescue stand persistence during periods of drought and high temperature stress and under grazing in the southern USA (Read & Camp, 1986; West, 1994). Increased grass persistence due to endophyte presence is especially beneficial for soil stabilization and profitable livestock production on marginal soils. Therefore, the driving forces for novel endophyte technology is the maintenance of plant persistence and the deletion of animal toxicosis. Endophyte technology has rapidly and easily been applied to improving fescues and ryegrasses used for turf because of less importance in deleting animal toxins (Funk, 1994).

## APPROACHES

Approaches to improving grass-endophyte associations for forage use can take three forms. One approach is to genetically tranform the endophyte by disrupting the synthesis or activity of enzymes responsible for synthesis of toxic alkaloids (Schardl, 1994). The determining and rate-limiting enzyme for ergot alkaloid synthesis is dimethylallyltryptophan synthase; however, progress lags in identifying the gene responsible and understanding its structure and regulation. Tsai et al. (1992) demonstrated that *N. coenophialum* can be transformed, but the optimal method of genetically engineering an endophyte not to produce ergot alkaloids has not been developed.

Another approach is to breed the host plant population for suppression of toxin production. Adcock et al. (1997) reported heritabilities for low ergot alkaloid concentration of 0.56 to 0.91. Reductions in toxin production have occurred without suppression of endophyte growth in the plant (Adcock et al., 1997), suggesting that the beneficial aspects of symbiosis would prevail. Therefore, this approach does have potential to reduce toxin concentration in the forage; however, it would not necessarily eliminate toxin production. In theory, there is the possibility that higher toxin production may occur with genetic drift of the host population or if foreign pollen invades a population and a selection pressure favors increase of high-ergot-alkaloid plants (Hill, 1993). Preventing seed production in forage stands through sound grazing and haying management would prevent the ingress of genes favoring high toxin production.

A third approach, the subject of this paper, is to select endophytes that naturally do not produce animal toxins, and then to transfer them into improved cultivars. This approach assumes that enough genetic variability exists within *N. coenophialum* and *N. lolii* to allow selection of such a strain. The advantage of endophyte selection over host breeding is that the former relies solely on endophyte genetics to control toxin production in the neosymbiotum. This means that the desirable alkaloid profile of the endophyte can be transferred into many host populations. Furthermore, an endophyte strain that is stable in its inability to produce toxins will theoretically maintain the nontoxic phenotype despite environmental variation and genetic drift in the host population. Endophyte selection allows fairly rapid development of new populations via inoculation of seedlings and subsequent seed production, as opposed to multigenerational recurrent selection or backcrossing programs involved in breeding the host population for toxin suppression. Drawbacks to the endophyte selection approach are the necessity to assemble and screen large numbers of diverse germplasm for alkaloid production and to confirm beneficial traits; in addition, there is the potential for poor compatibility with the target host population. Ultimately, the most successful means of eliminating fescue toxicosis potential may be via some combination of plant selection and fungal selection or engineering, or in the introduction of several strains of fungus into a population.

## PERENNIAL RYEGRASS

Fletcher and Easton (1997) described pioneering work in New Zealand in which novel endophyte technology was first used to correct an antiquality disorder in sheep grazing perennial ryegrass. They acquired E+ ryegrass germplasm from existing and new collections (Latch, 1989), then used high performance liquid chromatography (HPLC) to screen *N. lolii*-ryegrass associations for desired alkaloid profile. Strains were found that lacked lolitrem B but possessed peramine, an alkaloid conferring resistance to Argentine stem weevil (Rowan et al., 1986).

A candidate endophyte was transferred into an E– host population via seedling inoculation and then grown in pastures. The desired alkaloid profile prevailed in the grass, with the result that sheep showed no ryegrass staggers, and resistance to Argentine stem weevil was retained. 'Grasslands Greenstone' and 'Grasslands Pacific' are two cultivars that are marketed in New Zealand under the Endosafe (AgResearch, Palmerston North, New Zealand) trademark, indicating a beneficial endophyte (Rolston, 1993). They later also screened for the absence of ergovaline, the main representative of ergot alkaloids involved in ergot toxicosis.

## TALL FESCUE

We followed a regime at the University of Arkansas similar to that of Fletcher and Easton (1997) to eliminate fescue toxicosis while maintaining pest resistance; however, there is no particular insect that threatens stand persistence in tall fescue in the USA, like the Argentine stem weevil does in New Zealand. Our emphasis is on selecting endophytes that promote persistence under environmental stress on marginal soils of the Ozark Plateau region of north Arkansas and south Missouri. The sequence starts with screening germplasm and culminates with testing animal performance and stand persistence under grazing.

### Germplasm Acquisition and Endophyte Detection

Seed samples from the USDA Plant Introduction (PI) collection of tall fescue were obtained from the station at Pullman, WA. Seeds from each of 565 accessions were microscopically examined (Holder et al., 1994) for endophyte presence (new accessions have been examined as they became available). Only 79 accessions contained endophyte in the seed at a mean infection frequency of 67%. Presence in the seed does not assure viability of endophyte. Therefore, to confirm endophyte viability, seeds of each accession containing any endophyte were planted in flats in the greenhouse, and 10-wk-old seedlings were microscopically examined for endophyte presence. Only 55 of those 79 accessions contained viable endophytes, and those were at a mean infection frequency of 41%. There was apparently a massive loss of potentially beneficial endophytes from the tall fescue collection because seed storage and multiplication procedures deemed appropriate for the seed were used. These procedures were inadequate for endophyte preservation, which limits the value of old seed collections as sources of beneficial endophytes.

Collections of new accessions were made to acquire freshly harvested seed with viable endophytes. First, seed was collected from numerous randomly selected locations in the Ozark Plateau in pastures and roadsides and from pastures whose owners reported lack of fescue toxicosis symptoms in cattle. Then, to broaden the germplasm base beyond the 'Kentucky-31' tall fescue grown locally and to partially replenish the USDA-PI collection of endophytes from

*Festuca* sp., seed also was collected in Morocco, Tunisia, Spain, and France (West et al., 1993). These locations were chosen because of interest in obtaining endophytes from plants adapted to dry, hot summers. We identified the species and subspecies of collected *Festuca* to gain information on the compatibility of these endophytes when later introduced into tall fescue cultivars (McConnell et al., 1993a).

## Chemical Screen

The next step in our selection regime was to conduct an extensive analysis of the ergopeptine alkaloid, ergovaline (Rottinghaus et al., 1991), on leaf samples of greenhouse-grown accessions containing viable endophyte. This allowed us to reduce the number of candidate plants to a small enough number to perform a more intensive reanalysis of ergovaline, plus analyses of the known insect-deterring alkaloids, *N*-acetyl loline and *N*-formyl loline (Bush et al., 1982; Johnson et al., 1985), and peramine (Tapper et al.,1989). The object was to select plants whose ergovaline levels were below the detection limit (detection level 50 ng $g^{-1}$), but that contained measurable amounts of lolines and peramine. The criterion for selecting accessions for further study and endophyte isolation was the complete lack of chromatogram peaks that were suspected to represent any ergopeptine alkaloids. Ergovaline was chosen as the marker livestock toxin because it is the predominant ergopeptine alkaloid present in E+ tall fescue (Lyons et al., 1986). In addition, lysergic acid amide (ergine), the primary nonergopeptine ergot alkaloid found in E+ tall fescue (Powell & Petroski, 1989), can potentially contribute to the causes of fescue toxicosis (McCollough et al., 1994). Accessions that contained no detectable ergovaline were also devoid of lysergic acid amide.

Among the local collections, no low-ergovaline plants were found, which agrees with New Zealand reports on locally collected perennial ryegrass (Fletcher & Easton, 1997). Plants from the PI collection and our Mediterranean collections showing ergovaline concentrations below the detection limit were advanced to the second stage of resampling and reanalysis. There were no significant correlations among ergovaline, peramine, or loline concentrations, indicating that these alkaloids can be screened independently. Improvements to our chromatography procedure lowered the detection limit to 10 ng $g^{-1}$ (Moubarak et al., 1996). Greater HPLC resolution plus further plant analysis from field and greenhouse plantings of seedhead, stem, and crown plus pseudostem resulted in rejection of many earlier candidate accessions because of the detection of ergovaline or other ergot alkaloid peaks in the samples.

## Insect Screen

The objective of insect screening is to use bioassays to verify that plant-endophyte selection based on alkaloid profile translates into actual pest resistance even in the absence of ergot alkaloids. We developed E– clones of the selected E+, ergovaline-deficient wild plants through multiple drenchings with a systemic fungicide (Bacon & White, 1994). The E+ and E– plants were subjected to non-choice feeding tests of bird-cherry oat aphid (*Rhopalosiphum padi* L.) and greenbug (*Schizaphus graminum* Rondani), insects that are deterred by *Neotyphodium* endophytes in grasses (Siegel et al., 1990). Results indicated a high level of host resistance in both E+ and E– plants, thereby making it difficult to assess the ability of the endemic endophytes to enhance insect resistance. This procedure was time consuming, labor intensive, and fairly inconclusive; therefore, pest bioassays are more efficient when performed on the newly inoculated cultivar to verify the improvement in pest resistance.

## Endophyte Inoculation

Once a limited number of E+ plants are identified that possess the desired alkaloid profile, the endophytes are isolated on potato dextrose agar (Bacon & White, 1994). The isolates are subcultured on cellophane film covering proline-glutamic acid agar (Siegel, 1994, personal communication). Cellophane allows the diffusion of nutrients to the mycelia, while permitting the harvest of minute amounts of mycelia from the colony growing-front without picking up agar fragments. It is advisable during the endophyte culturing phase to subculture the isolates for *Neotyphodium* species identification if there is doubt as to its origin or host species.

Perennial ryegrass and tall fescue are largely self-infertile and therefore depend on cross pollination for seed production. Synthetic cultivars comprise genetically unique seeds derived from random mating among a number of selected clonal parents (Pohlman & Sleper, 1995). It is therefore critical to ensure that the host population is adequately represented among the successfully infected plants when performing endophyte inoculations. A seemingly simple way to achieve this is to inoculate the clonal parents and then allow the normal process of random mating and seed transmission to carry the endophyte into succeeding generations when increasing seed. This is effective provided that the number of clonal parents is small (a dozen or fewer) and if the endophyte strain is compatible with all or most of the clonal genotypes.

A major technical barrier to clone inoculation is the virtual impossibility of introducing endophytes into mature plants. Parrot (1994) described techniques of artificially infecting somatic embryos as a way to simulate the natural process by which seed-borne endophytes infect germinating embryos, thereby circumventing the natural resistance of mature plants to accepting endophytes. Somatic embryos derived from tissue culture provide numerous seedling-like tissues that are, in theory, genetically identical to the source plant. This is a potentially useful technique for studying host genotype x endophyte strain interactions (Hill et al., 1991). Regeneration of the newly infected somatic embryos from each parental clone of a synthetic cultivar to mature plants would allow one to reconstitute the cultivar with a novel endophyte. This technique, however, is limited by the following: (i) some plant genotypes are not conducive to callus formation (Hill et al., 1991), (ii) genotypes vary in their optimum culture medium for callus production and plant regeneration (McConnell et al., 1993b), and (iii) some genotypes are not compatible with particular strains of endophyte (Kearney et al.,

1991). There also is the possibility of somaclonal variation occurring in the regenerated plants, thereby compromising the effort to infect specific host genotypes.

Directly infecting excised apical meristems of mature plants with endophytes without going through tissue culture avoids the problems of unique medium requirements for each clone, long delays and tedium in tissue culturing and regenerating plants, and the possibility of somaclonal variation. Unfortunately, these attempts met with very low infection rates and high contamination with saprophytic and disease organisms (O'Sullivan & Latch, 1993; Ravel et al., 1994; Simpson et al., 1997).

The seedling-stab technique (Latch & Christensen, 1985) has become the most widely used technique for inoculating commercial cultivars. Details on the technique are given by Siegel and Bush (1994). Briefly, it involves germinating surface-sterilized breeder's seed on water agar for 5 d in the dark, stabbing a hole with an insect pin just above the coleoptilar node, and inserting a small mass of mycelia into the opening using the insect pin and forceps. The seedlings are placed back on the agar in the dark for 1 wk, then are transplanted to small pots under laboratory bench lights. Gradually the surviving seedlings are hardened to normal greenhouse conditions and checked for endophyte infection before transplanting to the field.

The advantages of seedling-stab over tissue culture techniques are that seedlings are quite receptive to foreign endophytes at <1 wk after germination, seedlings are much less prone to contamination problems, and large numbers can be inoculated at low cost and be ready for field transplanting within 6 mo. In effect, the inoculated seedlings become new parental clones. To regenerate the same cultivar, the recombined genes of the neosymbiotic seedlings must adequately represent the gene pool of the original clones. The main drawback of seedling-stab is that the resulting population of infected plants will exclude seedling genotypes that are incompatible with the new endophyte, thereby potentially changing the overall genetic makeup of the population. This is overcome simply by inoculating several hundred seedlings from a population in order to finally acquire at least 50 (H.S. Easton, 1993, personal communication) healthy, infected plants from that population for each endophyte strain to be introduced. Infecting smaller numbers of plants from the original population could result in reduced frequency of some characteristic genes of that population.

The seedling-stab technique precludes the ability to develop identical plant lines containing different endophyte strains since seeds of self-infertile grasses are genetically unique. Thus, studies of host × endophyte interactions using seedling stab have to involve identical host populations, not identical host genotypes. This approach, however, does reflect the practical field situation better than studies involving a few clones in that cultivars are genetically heterogeneous.

We have found in our laboratory that the tissue culture technique was slow, expensive, and suffered from extremely low success rate, high contamination, and high clonal incompatibility with the endophyte strains tested. It also is not practical to infect and regenerate clones for cultivars having a large number of parental clones. The seedling-stab technique was easily adopted in our laboratory, with success rates usually ranging from 20 to 60% depending on the endophyte strain. We have found only one strain out of 18 that had an unacceptably low success rate (West & Marlatt, 1996, unpublished data).

It is important when using the seedling stab technique to confirm that seed of the target cultivar is indeed E–. Heat treatment (Nott & Latch, 1993), aging, and ionization (Bagegni et al., 1990) can be used to rid seed of endemic endophytes.

## Seed Production and Evaluation

The success of infection should be checked at every stage of seed production such as before field transplanting, before anthesis, before bulking seed from individual plants in the pollination block, and during seed increase of the next generation. Completely incompatible endophytes simply do not grow in the new seedlings; however, some endophytes of low compatibility may infect the seedling, but die out after several months, or exhibit only partial transmission to the panicle and seed (Wilson & Easton, 1997). Essentially 100% seed transmission of endophyte is our criterion for full compatibility and is required for novel endophytes to have commercial viability.

It is equally important to confirm at the multiple steps of seed production that the desirable alkaloid profile is maintained in the inoculated plants and their progeny. Lack of ergovaline or total ergot alkaloids should be tested in plants before transplanting to the field pollination blocks, or at the latest, after transplanting but before anthesis. Toxin-producing plants discovered before anthesis can be used as sources of pollen to help maintain the cultivar gene pool (since endophytes are not paternally transmitted), but their seedheads should be clipped to prevent seed production. Another precaution when producing seed of a cultivar with a novel endophyte is to confirm that the plants are not at variance with the described cultivar. Progeny testing is used in New Zealand to assure cultivar integrity (Rolston, 1993). The identity of the endophyte strain in its neosymbiosis can be ascertained as matching the identity of the inoculum by reisolating the endophyte from tillers of the inoculated plant or from progeny seed and by comparing its DNA markers with those of the inoculum using RFLP or RAPD techniques.

We found that the first new seed (Syn2 generation, if the breeder's seed used for inoculation is considered Syn1) produced from 50 inoculated plants produced around 500 g of seed. This is enough seed for assays of insect and nematode resistance and greenhouse drought trials. We have determined that these traits are more conveniently and meaningfully tested with Syn2 seed of the neosymbiosis than with the original wild clones that provided the endophytes.

Most of the remaining seed is used to plant a larger seed field and bulk up larger quantities. This seed (Syn3) is used to establish pastures for confirming desirable animal performance and plant persistence. If larger amounts of Syn2 are produced and low seeding rates are used to carefully establish small paddocks, then sheep can be used for initial animal evaluation. The problem with field testing on a pasture scale is that it takes a long time to determine persistence. Because breeders wish to release cultivars as soon as possible, predictors of field persistence need to be developed to allow assessment at an earlier stage in the cultivar development program. The presence of peramine in perennial ryegrass and the resulting host resistance to Argentine stem weevil is apparently adequate to presume good field persistence in New Zealand (Fletcher & Easton, 1997). With tall fescue in the southern USA, however, persistence is determined by interacting factors of drought stress, grazing stress, and possibly nematode parasitism (West & Gwinn, 1993), for which no single, early indicator currently exists. Pasture evaluation should therefore be conducted on soils where an E– check population is expected to persist poorly.

A major concern in the commercial seed industry about using novel endophytes is the added cost of obtaining, testing, and introducing beneficial endophytes, but even more seriously, of ensuring the viability of the new endophyte as the seed passes through the merchandising channels. More costly packaging, storage, and marketing procedures will need to be adopted to assure that the benefits of the new endophyte are maintained with the seed. The turf industry has already made much use of novel endophytes, but the economic viability of using novel endophytes in commercial forage cultivars is yet to be determined.

## ACKNOWLEDGMENT

We appreciate the many helpful discussions about novel endophyte technology with H.S. Easton, L.R. Fletcher, N.S. Hill, G.C.M Latch, and M.R. Siegel. Many thanks are directed to S. Saidi, A. Benjeddi, J.M. Oliviera, and G. Charmet for local assistance in plant collection.

## REFERENCES

Adcock, R.A., N.S. Hill, J.H. Bouton, H.R. Boerma, and G.O. Ware. 1997. Symbiont regulation and reducing ergot alkaloid concentration by breeding endophyte-infected tall fescue. J. Chem. Ecol. 23:691–704.

Bacon, C.W., and J.F. White, Jr. 1994. Stains, media, and procedures for analyzing endophytes. p. 47–56. *In* C.W. Bacon and J.F. White, Jr. (ed.) Biotechnology of endophytic fungi of grasses. CRC Press, Boca Raton, FL.

Bagegni, A.M., D.A. Sleper, H.D. Kerr, and J.S. Morris. 1990. Viability of *Acremonium coenophialum* in tall fescue seed after ionizing radiation treatments. Crop Sci. 30:1272–1275.

Bush, L.P., P.L. Cornelius, R.C. Buckner, D.R. Varney, R.A. Chapman, B.P. Burris II, C.W. Kennedy, T.A. Jones, and M.J. Saunders. 1982. Association of *N*-acetyl and *N*-formyl loline with *Epichloe typhina* in tall fescue. Crop Sci. 22:941–943.

Coley, A.B., H.A. Fribourg, M.R. Pelton, and K.D. Gwinn. 1995. Effects of tall fescue endophyte infestation on relative abundance of small mammals. J. Environ. Qual. 24:472–475.

Fletcher, L.R., and H.S. Easton. 1997. The evaluation and use of endophytes for pasture improvement. p. 209–227. *In* C.W. Bacon (ed.) Proc. 3rd Int. Symp. *Neotyphodium*–Grass Interactions, Athens, GA. 28–31 May 1997. Plenum Press, New York.

Funk, C.R. 1994. Role of endophytes in grasses used for turf and soil conservation. p. 201–210. *In* C.W. Bacon and J.F. White, Jr. (ed.) Biotechnology of endophytic fungi of grasses. CRC Press, Boca Raton, FL.

Garner, G.B., G.E. Rottinghaus, C.N Cornell, and H. Testereci. 1993. Chemistry of compounds associated with endophyte–grass interaction: Ergovaline- and ergopeptine-related alkaloids. Agric. Ecosystems Environ. 44:65–80.

Gwinn, K.D., and A.M. Gavin. 1992. Relationship between endophyte infection level of tall fescue seed lots and *Rhizoctonia zeae* seedling disease. Plant Dis. 76:911–914.

Hill, N.S. 1993. Physiology of plant-endophyte interactions: Implications and use of endophytes in plant breeding. p. 161–169. *In* D.E. Hume et al. (ed.) Proc. 2nd Int. Symp. on *Acremonium*–Grass interactions: Plenary papers. AgResearch, Grassl. Res. Ctr., Palmerson North, New Zealand.

Hill, N.S., W.A. Parrott, and D.D. Pope. 1991. Ergopeptine alkaloid production by endophytes in a common tall fescue genotype. Crop Sci. 31:1545–1547.

Holder, T.L., C.P. West, K.E. Turner, M.E. McConnell, and E.L. Piper. 1994. Incidence and viability of *Acremonium* endophytes in tall fescue and meadow fescue plant introductions. Crop Sci. 34:252–254.

Hoveland, C.S. 1993. Importance and economic significance of the *Acremonium* endophyte to performance of animals and grass plant. Agric. Ecosystems Environ. 44:3–12.

Johnson, M.C., D.L. Dahlman, M.R. Siegel, L.P. Bush, G.C.M. Latch, D.A. Potter, and D.R. Varney. 1985. Insect feeding deterrents in endophyte-infected tall fescue. Appl. Environ. Microbiol. 49:568–571.

Kearney, J.F., W.A. Parrott, and N.S. Hill. 1991. Infection of somatic embryos of tall fescue with *Acremonium coenophialum*. Crop Sci. 31:979–984.

Latch, G.C.M. 1989. Plant improvement using endophytic fungi. p. 345–346. *In* Proc. XVI Int. Grassl. Cong. Nice, France. 4–11 Oct. 1998. Assoc. Française pou la Production Fourragère and Institut National de la Recherche Agronomique, Versaille, France.

Latch, G.C.M., and M.J. Christensen. 1985. Artificial infection of grasses with endophytes. Ann. Appl. Bot. 107:17–24.

Lyons, R.C., R.D. Plattner, and C.W. Bacon. 1986. Occurrence of peptide and clavine ergot alkaloids in tall fescue grass. Science (Washington, DC) 232:487–489.

McCollough S.F., E.L. Piper, Z.B. Johnson, R.J. Petroski, and M. Flieger. 1994. Effect of tall fescue ergot alkaloids on peripheral blood flow and serum prolactin in steers. J. Anim. Sci. 72:(Suppl. 1)144.

McConnell, M.E., C.P. West, and D.A. Sleper. 1993a. Ploidy level and identification of fescue species collected in the western Mediterranean region. p. 191. *In* Agronomy abstracts. ASA, Madison, WI.

McConnell, M.E., C.P. West, and D.A. Sleper. 1993b. Selection of media for callus production in fescue genotypes. Ark. Farm Res. 42(6):12.

Moubarak, A.S., E.L. Piper, Z.B. Johnson, and M. Flieger. 1996. HPLC Method for detection of ergotamine, ergosine, and ergine after intravenous injection of a single dose. J. Agric. Food Chem. 44:146–148.

Nott, H.M., and G.C.M. Latch. 1993. A simple method of killing endophyte in ryegrass seed. p. 14–15. *In* D.E. Hume et al. (ed.) Proc. 2nd Int. Symp. *Acremonium*–Grass Interactions, Palmerston North, New Zealand. 4–6 Feb. 1993. AgResearch, Grasslands Res. Ctr., Palmerston North, New Zealand.

O'Sullivan, B.D., and G.C.M. Latch. 1993. Infection of plantlets, derived from ryegrass and tall fescue meristems, with *Acremonium* endophytes. p. 16–17. *In* D.E. Hume et al. (ed.) Proc. 2nd Int. Symp. *Acremonium*–Grass Interactions. AgResearch, Grasslands Res. Ctr., Palmerston North, New Zealand.

Parrott, W.A. 1994. *In vitro* approaches for the study of *Acremonium-Festuca* biology. p. 37–46. *In* C.W. Bacon and J.F. White, Jr. (ed.) Biotechnology of endophytic fungi of grasses. CRC Press, Boca Raton, FL.

Pohlman, J.M., and D.A. Sleper. 1995. Breeding field crops. Iowa State Univ. Press, Ames.

Powell, R.G., and R.J. Petroski. 1989. Progress in chemistry of endophyte-infected tall fescue. p. 43–48. *In* Proc. Tall Fescue Toxicosis Workshop, SRIEG-37, Atlanta, GA.13–14 Nov. 1989.

Prestidge, R.A. 1993. Causes and control of perennial ryegrass staggers in New Zealand. Agric. Ecosyst. Environ. 44:283–300.

Ravel, C., D. Wartelle, and G. Charmet. 1994. Artificial infection of tillers from perennial ryegrass mature plants with *Acremonium* endophytes. p. 123–125. *In* K. Krohn et al. (ed.) Int. Conf. on Harmful Beneficial Microorganisms Grassl. Pastures Turf., Paderborn Germany. 4–6 Oct. 1994. Int. Organization for Biol. and Integrated Control of Noxious Animals and Plants, Montfavet, France.

Read, J.C., and B.J. Camp. 1986. The effect of the fungal endophyte *Acremonium coenophialum* in tall fescue on animal performance, toxicity, and stand maintenance. Agron. J. 78:848–850.

Rolston, M.P. 1993. Use of endophyte in plant breeding and the commercial release of new endophyte-grass associations. p. 171–174. *In* D.E. Hume et al. (ed.) Proc. 2nd Int. Symp. on *Acremonium*–Grass Interactions: Plenary papers, Palmerson North, New Zealand. 4–6 Feb. 1998. AgResearch, Grassl. Res. Ctr., Palmerson North, New Zealand.

Rottinghaus, G.E., G.B. Garner, C.N Cornell, and J.L. Ellis. 1991. HPLC method for quantitating ergovaline in endophyte-infested tall fescue: Seasonal variation of ergovaline levels in stems with leaf sheaths, leaf blades, and seed heads. J. Agric. Food Chem. 39:112–115.

Rowan, D.D. 1993. Lolitrems, peramine and paxilline: mycotoxins of the ryegrass/endophyte interaction. Agric. Ecosyst. Environ. 44:103–122.

Rowan, D.D., M.B. Hunt, and D.L. Gaynor. 1986. Peramine, a novel insect feeding deterrent from ryegrass infected with the endophyte *Acremonium loliae*. J. Chem. Soc. Chem. Commun. 935–936.

Schardl, C.L. 1994. Molecular and genetic methodologies and transformation of grass endophytes. p. 151–165. *In* C.W. Bacon and J.F. White, Jr. (ed.) Biotechnology of endophytic fungi of grasses. CRC Press, Boca Raton, FL.

Siegel, M.R. 1993. *Acremonium* endophytes: Our current state of knowledge and future directions for research. Agric. Ecosyst. Environ. 44:301–321.

Siegel, M.R., and L.P. Bush. 1994. Importance of endophytes in forage grasses, a statement of problems and selection of endophytes. p. 135–150. *In* C.W. Bacon and J.F. White, Jr. (ed.) Biotechnology of endophytic fungi of grasses. CRC Press, Boca Raton, FL.

Siegel, M.R., L.P. Bush, N.N. Fannin, G.C.M. Latch, D.D. Rowan, and B.A. Tapper. 1990. Occurrence and insecticidal activity of alkaloids in endophyte-infected plants. J. Chem. Ecol. 16:3301–3315.

Simpson, W.R., M.J. Christensen, and D.E. Hume. 1997. An appraisal of the use of axillary buds of grasses as clonal material for inoculation with *Neotyphodium* endophytes. p. 275–277. *In* C.W. Bacon (ed.) Proc. 3rd Int. Symp. *Neotyphodium*–Grass Interactions, Athens, GA. 28–31 May 1997. Plenum Press, New York.

Tapper, B.A., D.D., Rowan, and G.C.M. Latch. 1989. Detection and measurement of the alkaloid peramine in endophyte-infected grasses. J. Chromat. 463:133–138.

Tsai, H.-F., M.R. Siegel, and C.L. Schardl. 1992. Transformation of *Acremonium coenophialum*, a protective fungal symbiont of the grass *Festuca arundinacea*. Curr. Genet. 22:399–406.

West, C.P. 1994. Physiology and drought tolerance of endophyte-infected grasses. p. 87–99. *In* C.W. Bacon and J.F. White, Jr. (ed.) Biotechnology of endophytic fungi of grasses. CRC Press, Boca Raton, FL.

West, C.P., and K.D. Gwinn. 1993. Role of *Acremonium* in drought, pest, and disease tolerances of grasses. p. 131–140. *In* D.E. Hume et al. (ed.) Proc. 2nd Int. Symp. on *Acremonium*–Grass Interactions: Plenary papers, Palmerson North, New Zealand. 4–6 Feb. 1993. AgResearch, Grassl. Res. Ctr. Palmerson North, New Zealand.

West, C.P., D.A. Sleper, and S. Saidi. 1993. Collection of endophyte-infected fescue species in Morocco, Spain, and France. p. 196. *In* Agronomy abstracts. ASA, Madison, WI.

Wilson, S.M., and H.S. Easton. 1997. Seed transmission of an exotic endophyte in tall fescue. p. 281–283. *In* C.W. Bacon (ed.) Proc. 3rd Int. Symp. *Neotyphodium*–Grass Interactions, Athens, GA. 28–31 May 1997. Plenum Press, New York.

# 9 Somatic Embryogenesis: Forage Improvement using Synthetic Seeds and Plant Transformation

**Bryan D. McKersie and Steve R. Bowley**

*Plant Biotechnology Division*
*Department of Plant Agriculture*
*University of Guelph*
*Guelph, Ontario, Canada*

## ABSTRACT

Somatic embryogenesis is fundamental to the genetic improvement of forage legumes for plant transformation and for propagation by synthetic seeds. The trait is sexually inherited in alfalfa as two independent, complementary, dominant nuclear genes. Populations that have different genetic backgrounds, that contain a high proportion of embryogenic individuals and that have good field performance have been created. To evaluate the potential of transformation technology to improve winter survival in alfalfa, we created transgenic alfalfa plants expressing a cDNA for a Mn-superoxide dismutase (SOD). In replicated field plots, all SOD transgenic plants had much higher survival after two winters than the non-transgenic genotype. Releasing these or other transgenic alfalfa plants as commercial varieties requires special consideration because of autotetraploid inheritance and in-breeding depression during seed production. Several alternative seed production methods are described that might produce transgenic varieties from a few transgenic parents. One model is to backcross the transgenic plant into different populations and then identify new parents for the synthetic variety from those populations. Synthetic seeds might be used to produce a double-cross hybrid or a reduced generation synthetic. Alternatively, one or two transgenic plants might be included as parents in conventional synthetic varieties and herbicide resistance used to select those progeny containing the desired transgenes. This last strategy requires the least laboratory labor and over 80% of the forage production plants probably might contain the transgene.

## GENETICS OF EMBRYOGENESIS

Somatic embryogenesis is fundamental to the genetic improvement of forage legumes using biotechnology. The technique to form somatic embryos using only induction and regeneration steps was first described in alfalfa (*Medicago sativa* L.) 25 yr ago (Saunders & Bingham, 1972). This relatively simple procedure has been modified to enable transformation of alfalfa using *Agrobacterium tumefaciens* and a more complex multiple-step procedure has been developed to make synthetic seeds. Both these techniques will be discussed in more detail in the following sections.

---


*Molecular and Cellular Technologies for Forage Improvement*. CSSA Special Publication no. 26.

First, it must be realized that only certain alfalfa plants will respond in culture to form somatic embryos, or in other words, the ability of alfalfa to form somatic embryos in tissue culture is genotype dependent (Saunders & Bingham, 1972; Brown & Atanassov, 1985). Although significant interactions have been reported among genotypes and culture protocols (Meijer & Brown, 1985; Seitz-Kris & Bingham, 1988), an embryogenic genotype is an essential prerequisite to the application of many techniques in biotechnology. Forage legumes are one of the few groups of plants in which the embryogenic response has been studied from a genetic point of view (reviewed by Henry et al., 1994). In an extensive survey of *Medicago* germplasm, the best regenerating cultivars of alfalfa were found to possess a genetic background that included *M. falcata* (Brown & Atanassov, 1985). One selection from this screening, the genotype A70-34 (synonym RL-34) from the cultivar Rangelander, is highly embryogenic and genetically stable during in vitro culture (McKersie et al., 1989); however, those are its only redeeming agronomic traits because it is susceptible to most diseases and lacks the vigor of commercial alfalfa germplasm in field environments. The somatic embryogenesis trait is sexually heritable in a wide range of plants including alfalfa and is conditioned by nuclear genes (Henry et al., 1994). Crosses made among embryogenic and nonembryogenic plants have shown that the somatic embryogenesis character in alfalfa is controlled by two independent, complementary, dominant genes (Reisch & Bingham, 1980; Hernandez-Fernandez & Christie, 1989; Kielly & Bowley, 1992; Crea et al., 1995).

In our experience, none of the commercial-quality germplasm adapted to our field environment would readily form somatic embryos at a high frequency. Therefore, a modified backcrossing program was established in which RL-34 was used as the source of the embryogenesis genes and this parental plant was crossed with several individual plants selected from varieties adapted to Northeastern North America. $F_1$ progeny were selected for the ability to form somatic embryos, and these were crossed to other adapted plants. The process was repeated for two cycles and the plants were evaluated in the field for their relative forage and seed yields and other performance criteria (Bowley et al., 1993). The results indicated that regeneration by somatic embryogenesis and agronomic performance such as forage yield, seed yield, and persistence could be combined. The populations have been backcrossed for an additional two cycles and are now being evaluated for combining ability.

We currently have seven populations created in this way that have different genetic backgrounds, that contain a high proportion of embryogenic individuals and that have good field performance in terms of forage yield, seed yield, stress tolerance, and disease tolerance. This is one of the few examples of a plant breeding program that has purposefully created commercial populations that are amenable to manipulation using biotechnology.

## TWO STEP TISSUE CULTURE SYSTEM

Our alfalfa tissue culture systems are based on the two-step procedure that was originally developed by Saunders and Bingham (1972), but modified in different ways to either enhance embryogenesis in specific genotypes or to achieve specific types of embryo development. In the basic procedure, petiole explants are harvested and surface sterilized (Fig. 9–1). The explants are then placed on a solid agar medium containing either SH (Schenk & Hildebrandt, 1972) or B5 (Gamborg et al., 1968) salts, an auxin (usually 2,4-D, 2,4-dichlorophenoxy acetic acid), and a cytokinin (usually kinetin). On this induction medium (Step 1), the cells in the vascular cambium divide producing a callus, whereas some of those in the subepidermal layer divide producing somatic embryos (Wenzel & Brown, 1991). The proembryogenic structures develop into globular somatic embryos and regeneration is greatly facilitated if the culture is moved to a growth-regulator-free BOi2Y (Bingham et al., 1975) medium (Step 2). The embryos develop and precociously germinate into rooted seedlings that can then be removed from culture and transplanted into pots in the greenhouse or growth cabinet. Numerous modifications of the organic and inorganic components of the induction medium have been made, but the presence of 2,4-D or a similar strong auxin is one of the critical factors. Some auxins, including indole-3-acetic acid (IAA) and indole-3-

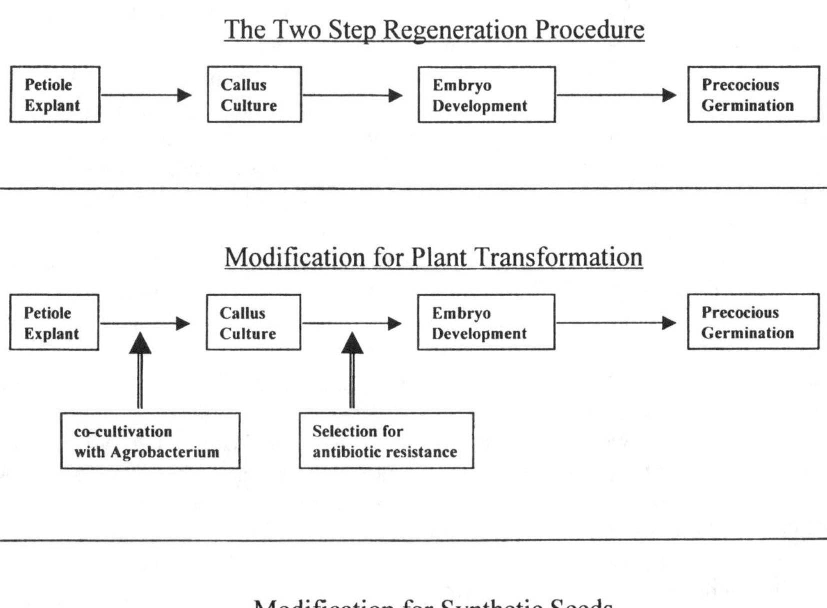


Fig. 9–1. Modifications of the basic two step alfalfa tissue culture system used for plant transformation and for synthetic seed production.

butyric acid (IBA), are ineffective and neither callus nor somatic embryos develop (McKersie & Brown, 1996). Other auxins including α-napthaleneacetic acid (NAA) and chlorophenoxyacetic acid (CPA) cause callus proliferation but do not induce embryo formation.

## TRANSFORMATION OF ALFALFA

This two-step regeneration procedure can be easily modified to produce alfalfa plants that have been genetically engineered using *Agrobacterium* (Fig. 9–1). The plant genotype is one of the most critical factors in the successful transformation of alfalfa. The genotype must be not only capable of regenerating via somatic embryogenesis at a very high frequency but also susceptible to infection by *Agrobacterium*. Several of the plants from the breeding program described above satisfy these criteria and we did not find it necessary to extensively screen for susceptibility to *Agrobacterium* in these populations. The following procedure is being routinely used in our laboratory but there are numerous variations that others have used (Austin et al., 1995; Desgagnes et al., 1995; D'Halluin et al., 1990; Du et al., 1994; McKersie et al., 1993; Narvaez-Vasquez et al., 1992; Samac, 1995; Shahin et al., 1986; Spano et al., 1987; Thomas et al., 1994). The petiole explants are surface sterilized in 75% ethanol for 30 s and in 4% Ca hypochlorite for 20 min. Then, they are cut into 1-cm sections and briefly soaked in a log phase (OD600 = 1.0) culture of *Agrobacterium tumefaciens*. We use strain C58C1Rif pMP90, but other strains may be equally effective. The petioles and *Agrobacterium* are cocultivated in the dark at 25°C for 2 to 3 d on modified SH induction medium containing 100 μ*M* acetosyringinone. The petioles are then washed with half-strength MS salt solution to remove the *Agrobacterium* and plated on the same induction medium but including 500 mg $L^{-1}$ claforan and 50 mg $L^{-1}$ kanamycin (or other selection agent). The petioles are transferred to a fresh medium every 14 d because claforan degrades in the light and the dead plant tissues produce phenolics that leach into the medium.

After several weeks, the majority of the explants become necrotic and die in the presence of the selection agent. Small calli appear and somatic embryos form either on the callus or directly on the explant. Although the frequency of somatic embryo production varies among experiments, approximately 10% of our explants will form somatic embryos. Once initiated, the somatic embryos or callus may be transferred to either BOi2Y development medium or half-strength MS medium (Murashige & Skoog, 1962) usually lacking the selection agent. Our greatest numbers of transformants are obtained by transferring torpedo-shaped embryos without any attached callus directly to growth regulator-free, half-strength MS medium.

The regenerated seedlings are checked using PCR (polymerase chain reaction) for the transgenes before the plants are moved to the greenhouse. Using the transformation procedure described, very few of the several hundred regenerated seedlings produced in the last 2 yr have scored PCR-negative. When negative PCR results have been obtained, either all plants in a particular batch were negative, meaning that there had been an error in the selection procedure, or there were difficulties in the DNA extraction leading to false negatives. Although confirmed escapes have been rare, not all regenerated plants express the transgenes in the desired manner (see below).

## Transformation for Improved Winterhardiness

Winterhardiness is a complex trait involving tolerances to freezing, water deprivation, ice-encasement (severe anoxia), flooding (milder anoxia), and disease. The combination and severity of the stresses that crops must tolerate during winter varies from environment to environment and from year to year. Progress in the development of more winterhardy alfalfa plants has been slow. One of the reasons for the limited progress has been the difficulty in combining the many genes associated with yield, quality, disease and several stress tolerances. Another reason is the difficulty in quantifying the degree of winterhardiness. These limitations in our current methods prompted us to explore the feasibility of using a genetic engineering approach to improve winterhardiness, where genes encoding for improved winterhardiness could be introduced into good parental plants of a commercially accepted cultivar. A major challenge was the choice of the genes to be introduced or modified by genetic engineering. We decided to focus our efforts not on individual aspects of freezing, anoxia or desiccation tolerance but on the common elements among the stresses that cause winter injury. In other words, even though freezing, anoxia, and desiccation are recognized as being distinctly different types of environmental stresses, they have common physiological elements, one being oxidative stress. Our hypothesis was that if we could enhance the plant's tolerance of oxidative stress, then we would improve its ability to survive the multiple stresses associated with winter.

The physiological and biochemical evidence that links oxidative stress with freezing, anoxia and desiccation stress is only correlative, but four major experimental observations have been made. The injury symptoms that were observed in cellular membranes isolated from plants exposed to freezing temperatures, ice-encasement, desiccation, or the free radical generating herbicide paraquat were very similar (Borochov et al., 1987; Hetherington et al., 1988; Kendall et al., 1985; Senaratna et al., 1985). Secondly, the membranes from plants acclimated to low temperatures were more tolerant of in vitro free radical treatment than those from nonacclimated plants (Kendall & McKersie, 1989). Thirdly, as plants acclimated at low temperatures, they acquired coincidentally increased tolerance to freezing stress, ice-encasement stress and free-radical generating herbicides (Bridger et al., 1994; Hetherington et al., 1988). Finally, when alfalfa plants were subjected to a freezing stress, their antioxidant levels were depleted coincidently with the loss of viability (Schubert, 1994).

To test our hypothesis, we collaborated with Chris Bowler and Dirk Inzé at the Laboratorium voor Genetika, Rijksuniversiteit, Gent and Kathleen D'Halluin and Johan Botterman at Plant Genetic Systems, Gent. The focus of our attention was superoxide dismutase (SOD), an enzyme that catalyzes the dismutation of two superoxide molecules to hydrogen peroxide and oxygen (Bowler et al., 1992, 1994; Foyer et al., 1994; Scandalios, 1990, 1993). A Mn-SOD cDNA had been isolated from *Nicotiana plumbaginofolia* (Bowler et al., 1989). Chris Bowler

Table 9–1. Survival of alfalfa (*Medicago sativa* L.) expressing a Mn-superoxide dismutase transgene in field trails at Elora, Ontario, after transplanting in spring 1992.†

| Plant | Fall 1992 | Fall 1993 | Fall 1994 |
|---|---|---|---|
| | | % survival‡ | |
| RA3 | 97a | 42b | 17c |
| RA3-ChlSOD§ | 93a | 47b | 41b |
| RA3-MitSOD§ | 91a | 62a | 54a |

† Data summarized from McKersie et al., 1996.
‡ Means within a column followed by the same letter are not significantly different at the 0.05 probability level using a Duncan's multiple range test.
§ Mean of two independent transgenic plants.

constructed binary transformation vectors controlling the expression of the Mn-SOD cDNA with the CaMV35S promoter and targeting the Mn-SOD protein to either the mitochondria or chloroplasts (Bowler et al., 1991). These vectors were used to transform an alfalfa plant called RA3 using *Agrobacterium tumefaciens* (McKersie et al., 1993). This RA3 plant was used because when we did these transformations in 1988, this was one of the few alfalfa genotypes that would regenerate by somatic embryogenesis and which therefore could be transformed. The regenerated plants were screened for the presence of the transgene using PCR, the presence of a new SOD isozyme on native PAGE gels, the presence of unique T-DNA insertion using southern hybridization and inheritance of the transgenes to $F_1$ and $F_2$ progeny (McKersie et al., 1993). Total SOD activity in these plants was only modestly increased, in part because increased Mn-SOD activity seems to be accompanied by lower Cu–Zn-SOD activity in most transgenic plants (B.D. McKersie, 1995, unpublished data). The transgenic plants did not have increased tolerance of the herbicide paraquat in an excised leaf assay, but they were slightly more tolerant of the diphenyl ether herbicide, acifluorfen. They also tended to have higher shoot regrowth from crown and root tissues after freezing stress. The tolerance of freezing was heritable to the $F_1$ progeny; those plants that had inherited the Mn-SOD transgene also had less freezing injury compared with those progeny lacking the transgene across the temperature range between −8 and −14°C.

Our hypothesis that enhanced tolerance of oxidative stress provides increased tolerance to multiple environmental stresses was supported further by observations that these same plants have increased tolerance of water deprivation. When deprived of water in growth cabinets for 5 d, plants expressing the Mn-SOD cDNA tended to have smaller changes in the chlorophyll fluorescence ratio, Fv/Fm , and less electrolyte leakage from leaves (McKersie et al., 1996).

The most definitive test of our hypothesis has come from the evaluation of these plants in a field trial (McKersie et al., 1996). In 1992, 1 by 1.5 m replicated field plots were established using transplanted cuttings of four of these primary transgenic plants, two with mitochondrial targeting (pMitSOD) and two with chloroplast targeting (pChlSOD). As stated earlier, the RA3 plant that we had originally transformed was not adapted to our Ontario climate and its persistence in field plots was relatively low compared to commercial alfalfa varieties. Therefore, after 2 yr, only 17% of the original plants survived (Table 9–1); all

Table 9–2. Total seasonal herbage yield (g dry matter $m^{-2}$) of transgenic alfalfa (*Medicago sativa* L.) expressing a Mn-superoxide dismutase cDNA in field trials at Elora, Ontario, after transplanting in spring 1992.†

| Plant | 1992 | 1993 | 1994 |
|---|---|---|---|
| | | g $m^{-2}$‡ | |
| RA3 | 151b | 265c | 37b |
| RA3-ChlSOD§ | 289a | 559a | 140a |
| RA3-MitSOD§ | 228a | 432b | 164a |

† Data summarized from McKersie et al., 1996.
‡ Means within a column followed by the same letter are not significantly different at the 0.05 probability level using a Duncan's multiple range test.
§ Mean of two independent transgenic plants.

SOD transgenic plants had much higher survival than the nontransgenic RA3. This increased survival contributed to higher forage yields after both one and two winters (Table 9–2). The forage yields in the year of seeding (1992) also are interesting because those plants with the Mn-SOD transgene had numerically higher yields before the plants had experienced any winter stress. The reason for this apparent increase in vigor is unknown, but since this effect was not observed in controlled environment experiments, it possibly reflects tolerance of transplanting or water deprivation.

### Future Objectives in Genetic Engineering of Alfalfa

Given these very positive results from our first experiments, we concluded that it is possible to alter the agronomic performance of alfalfa using a genetic engineering approach. Our current objectives are to create genetic variability for persistence and forage quality—two factors that would improve the profitability of forage production in the dairy and beef industries of Ontario. To improve the persistence of alfalfa, we are introducing a series of genes related to oxidative stress tolerance and flooding tolerance. To improve forage quality, we are attempting to improve the energy content, digestibility, and ensiling quality of the herbage by manipulating the partitioning of carbohydrates. To be effective, a detailed knowledge of the biochemistry and physiology of these processes is required in order to identify candidate genes and we also require tissue-specific and environment-specific promoters that will enable transcriptional regulation of the transgene in a specific controlled manner that is different from the expression of the native gene and creates a different phenotype.

## INTRODUCTION OF TRANSGENIC PLANTS INTO SYNTHETIC VARIETIES

There are a number of hurdles that must be cleared before we can release transgenic alfalfa plants as commercial varieties. Of course, the first is regulatory approval, but the requirements in forage crops should not be any more stringent than those in vegetables, maize (*Zea mays* L.) or canola (*Brassica* sp). So we are optimistic that these requirements will remain reasonable. The second prob-

Table 9–3. Percentage of plants in a synthetic alfalfa variety carrying a single transgene increased using three generations of conventional seed production from selected parental plants.†

| Generation | Parents with transgene 100 | 50 | 10 |
|---|---|---|---|
| | % | | |
| Breeder | 75.0 | 43.8 | 9.8 |
| Foundation | 70.7 | 42.2 | 9.7 |
| Certified | 69.1 | 41.6 | 9.6 |

† The model assumes a single gene insertion, chromosome segregation and autotetraploid inheritance.

lem is the autotetraploid inheritance of transgenes in alfalfa, which basically means that it will not be possible to ensure that all individuals in the population carry the transgene. The third problem is the cross pollinating nature of alfalfa and in-breeding depression. To minimize in-breeding, our current alfalfa cultivars are produced using several (as many as 100 or more) parental plants multiplied through three or four generations of random mating.

The most obvious strategy to release transgenic plants as commercial synthetic varieties is to use the current seed production system and include the transgenic plants as parents, but unless all the parents are transgenic, the transgene would be at a relatively low frequency in the Certified seed (Table 9–3) . For example, if 10% of the parents were transformed with a single copy of the transgene, and three generations were used to produce Certified seed, <10% of the plants in the forage production field would contain the transgene. This is unlikely to produce a variety that is significantly altered for the trait conditioned by the transgene. A variation on this strategy would be to increase the number of copies of the T-DNA in the original parent; however, this would be effective only if copy number had no effect on phenotype; in other words, single T-DNA insertions would have to give the same phenotype as multiple T-DNA insertions. Only if all of the parents were transformed with a single copy of the transgene would the proportion of transgenic progeny in the Certified generation be sufficient to provide the transgenic phenotype at a commercially acceptable level (i.e., >50% of the plants expressing the desired trait).

There are additional complications with this strategy. First, transgene expression varies among independent transgenic plants even though they have incorporated the same transgene. An example of this effect is shown for transgenic alfalfa expressing the Mn-SOD transgene (Fig. 9–2). In this example, the genotype N4-4-2 was transformed with the pMitSOD vector (McKersie et al., 1993) and 25 plants were regenerated. The activity of MnSOD in leaf extracts was quantified after the SOD isozymes had been separated by native PAGE (McKersie et al., 1993) using a computer imaging system and calculating the activity in MnSOD as a percentage of the total SOD activity in the extract. Alfalfa leaves contain a MnSOD isozyme that in the nontransformed N4-4-2 plant accounted for 19% of the total SOD activity. Most of the transgenic plants had higher MnSOD activity as a consequence of the expression of the cDNA inserted in the T-DNA, but some plants did not and others had less activity. This com-

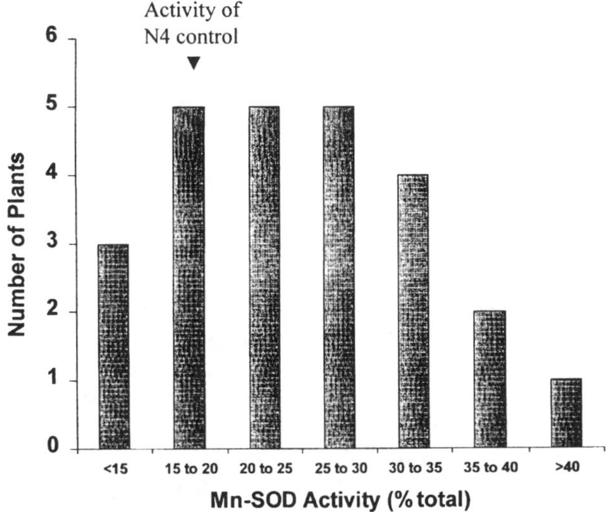


Fig. 9–2. Mn-superoxide dismutase (SOD) activity in the leaves of primary transgenic alfalfa plants expressing a Mn-SOD transgene and its native Mn-SOD gene.

monly observed variation in transgene expression is presumably due to the position of T-DNA insertion into alfalfa's chromosomes. This means from a practical point of view that perhaps 25 or as many as 100 transgenic plants need to be screened to find the one individual that has the desired expression and phenotype. If this screening had to be done for every parent of a synthetic cultivar, the cost of screening for complex phenotypes, such as forage quality or persistence, would make this genetic engineering approach prohibitively laborious.

There are several ways to address this complication. One could backcross the transgenic plant into different populations and then identify new parents for the synthetic variety from those populations, as has already been suggested (Micallef et al., 1995), but this adds a considerable amount of time and effort to the breeding program. We perceive an advantage to using a variety with known agronomic performance (e.g., yield and disease tolerance) and then to incrementally add transgenes for improved persistence or quality to that specific variety. This strategy would require that a traditional variety be created using nontransgenic parents that have the potential to form somatic embryos, and therefore have the potential to be transformed. If this was a double-cross hybrid, or a reduced generation synthetic that had few (e.g., four) parents, then we would need to create fewer transgenic parents. Alternatively, we might continue to use a large number of parents as in the current synthetic varieties, with only one or two additional parents being transgenic. Then, at each generation of seed increase, we might select only those progeny containing the transgene for production of the next generation. This could be accomplished relatively easily using a herbicide if the transgene also encoded herbicide resistance. The following sections explore these two options in more detail.

## MULTIPLE STEP TISSUE CULTURE PROCEDURE

Another application of somatic embryogenesis in forage improvement is the use of synthetic or artificial seed technology (Fig. 9–1). Our objective in developing synthetic seed technology was to devise a clonal propagation system that enabled multiplication of the donor plant and that also enabled the vegetative propagule to be stored for long periods of time. We had originally proposed that this might be a means to reduce the number of generations of seed increase and perhaps to produce reduced generation synthetic or hybrid varieties in alfalfa (McKersie et al., 1989; McKersie & Bowley, 1993). Somatic embryos seemed to be ideal for this purpose because we assumed that the somatic embryo could be forced to follow the same developmental program as the zygotic embryo in the seed leading to a dry quiescent structure. Before discussing how we might use this technology to introduce a transgenic variety into the market, we will first review the basic technology.

Synthetic seeds have been defined as somatic embryos engineered for use in the commercial propagation of plants (Gray & Purohit, 1991; Redenbaugh, 1993). Various forms have been envisioned over time. The first were simply hydrated somatic embryos. These had the particular advantage compared with cuttings of enabling rapid clonal multiplication of some plants, but the labor and therefore cost was high and the propagules were very delicate. This was partially overcome with the development of alginate coatings that encapsulated a single embryo enabling mechanized handling, but these hydrated encapsulated embryos could only be stored using low temperatures for a few weeks (Fujii et al., 1989, 1992). The capability of prolonged storage was achieved when the somatic embryos were dried to moisture contents <20% (McKersie et al., 1989).

When we began our research in the mid 1980s, the plant tissue culture systems used to produce synthetic seeds mimicked only the very first part of zygotic embryo development. The embryos did not accumulate reserves, they did not develop tolerance of drying, and they could not be stored. They simply precociously germinated into a seedling—a process that could be slowed by low temperature but not stopped. A diagrammatic representation of the plant tissue culture system that we subsequently developed in alfalfa is shown in Fig. 9–3.

The following is a brief description of the procedure; more detailed explanations are given elsewhere (McKersie & Bowley, 1993; McKersie & Brown, 1996). Petiole explants from greenhouse-grown plants are surface sterilized and cultured on induction medium as in the two-step procedure. The initial somatic embryos, which are only small dense cell clusters at this stage, are embedded in a callus mass of nondifferentiated cells. To liberate these proembryonic structures, and to stimulate the formation of more embryos, the callus is dispersed in a liquid medium to form a suspension culture. Usually, this liquid medium is modified B5 (Gamborg et al., 1968) containing 2,4-D but not kinetin; however, other salt combinations have been effective as well. After 7 d, the suspension is sieved and a 224–500 μm fraction is spread thinly on solid BOi2Y medium lacking 2,4-D (Bingham et al., 1975). On this medium, the embryos develop through morphological stages that appear to be globular, heart and torpedo. Once the majority of embryos reach the torpedo stage (7–10 d after sieving) they are transferred to an enriched BOi2Y medium containing a high level of sucrose, N and S to prevent precocious germination (Anandarajah & McKersie, 1990) and to enable deposition of storage reserves (Lai & McKersie, 1994). During this period, called maturation phase I in our experiments, the embryos rapidly accumulate fresh and dry weight, reaching 1 to 2 mg dry weight per embryo. To induce the acquisition of desiccation tolerance in maturation phase II, the somatic embryos are placed on a modified BOi2Y medium containing abscisic acid (ABA) for 3 d (Senaratna et al., 1989, 1990). Then, they are removed from the medium, washed to remove sugar and other nutrients, and dried. The standard method of drying is to place the somatic embryos in a sealed chamber over a saturated salt solution designed to give specific relative humidifies (Senaratna et al., 1989, 1990). Daily for 1 wk, the embryos are transferred to a progressively lower relative humidity chamber and finally they are air dried at ambient conditions. At this stage, the embryos have reached approximately 15% moisture and can be stored for a year or more with good viability. Subsequently, the embryos are germinated on moist filter paper (Lai et al., 1995). Although direct planting in the greenhouse is possible, direct field planting is at present impractical until coatings have been developed.

Synthetic seeds offer the opportunity to store genetically heterozygous plants with a unique gene combination that can not be maintained by conventional seed production because genetic recombination occurs at pollination. Transgenic plants also may be candidates for storage in this manner. Cryopreservation of desiccated somatic embryos may be another alternative for long term storage of valuable unique germplasm including transgenic plants (Yamada & Okumura, 1997).

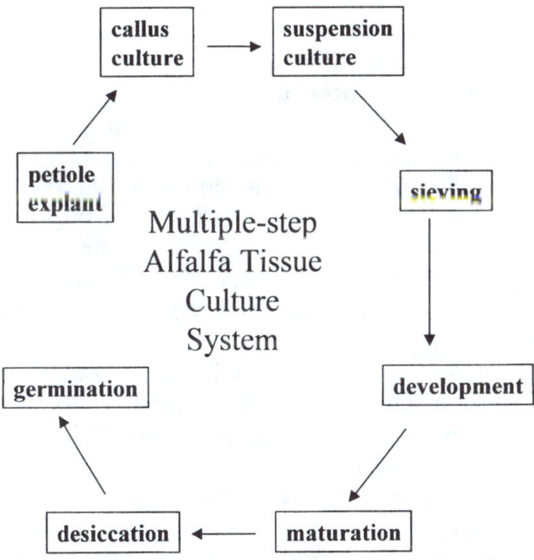


Fig. 9–3. The multiple step alfalfa tissue culture system used to produce synthetic seeds. The unique aspects of this culture system are the maturation phases designed to increase accumulation of storage reserves and acquisition of desiccation tolerance.

## USING SYNTHETIC SEEDS TO PROPAGATE TRANSGENIC PLANTS

One alternative that we have considered is the production of double-cross-hybrids or reduced-generation-synthetics to produce commercial quantities of transgenic seeds. These two methods differ in the pattern of crossing that is used to produce the commercial seed (Fig. 9–4). In both cases, synthetic seeds (or the more traditional vegetative propagation system, cuttings) might be used to replace Breeder seed. These could be used to establish Foundation seed production fields. The particular combination of plants in this seed production field might be in specific pairs, or might be as a random mixture of all parents, or another variation between these extremes. The result of this strategy would be that forage production fields would be seeded using the equivalent of Foundation seed, which is one generation earlier than the present system. This would reduce the amount of inbreeding that occurs during seed increase; however, its ability to produce a variety containing a high proportion of plants with a transgene would be similar to a traditional synthetic variety (compare Certified vs. Foundation in Table 9–3). Since relatively few parents are used in any of these seed production systems, relatively few transgenic parents would need to be created and evaluated.

## HERBICIDE SELECTION

From our perspective at the present time, the seed production system that will deliver the highest frequency of transgenic plants in a synthetic variety is one that uses only one or perhaps two transgenic parents, but applies a herbicide to select the transgenic progeny during both Foundation and Certified seed production (Fig. 9–5A). If a synthetic variety was produced from 10 parents, one being transgenic, and if a herbicide selection was applied to the Foundation seed production field, the proportion of plants in the forage production field (Certified seed) would be 72% (Table 9–4). If the herbicide was applied to both Foundation and Certified seed production fields, the proportion would be increased to almost 85%. With herbicide selection in both seed production fields, the proportion of plants in the Certified generation is not significantly different whether 10, 50, or 100% of the original parents were transgenic. This scheme would greatly reduce the amount of laboratory work necessary to create and to identify high performance transgenic plants.

Ideally, the herbicide should be effective as a selection agent both in cell cultures and in the field. There are several candidate genes for herbicide resistance but all are proprietary and none have been tested in alfalfa. We currently have tests in progress to determine the efficacy of using resistance to the phosphinothricin herbicides to select the transgenic progeny during seed production.

There are however complications and limitations to this scheme as well. The Foundation seed field would have to be overseeded about 10-fold to account for the selection, meaning that more Breeder seed would be required. Secondly, the transgenic parent would be a common parent for all surviving plants and

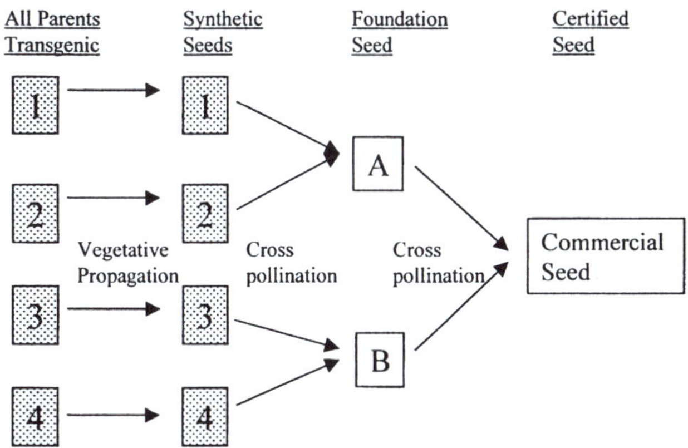


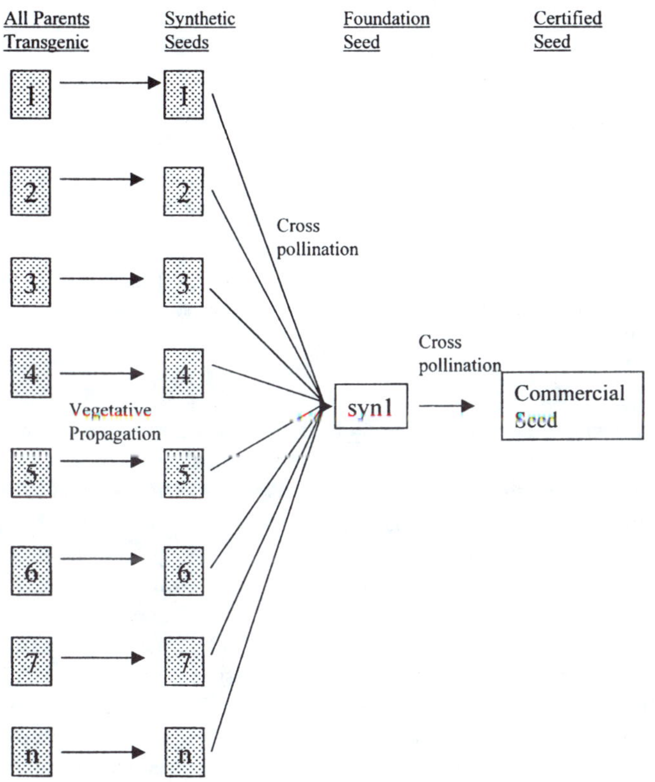


Fig. 9–4. Two seed production schemes that might be used to deliver an alfalfa variety as a (A) double crosss hybrid or (B) reduced generation synthetic assuming that all parents are transgenic.

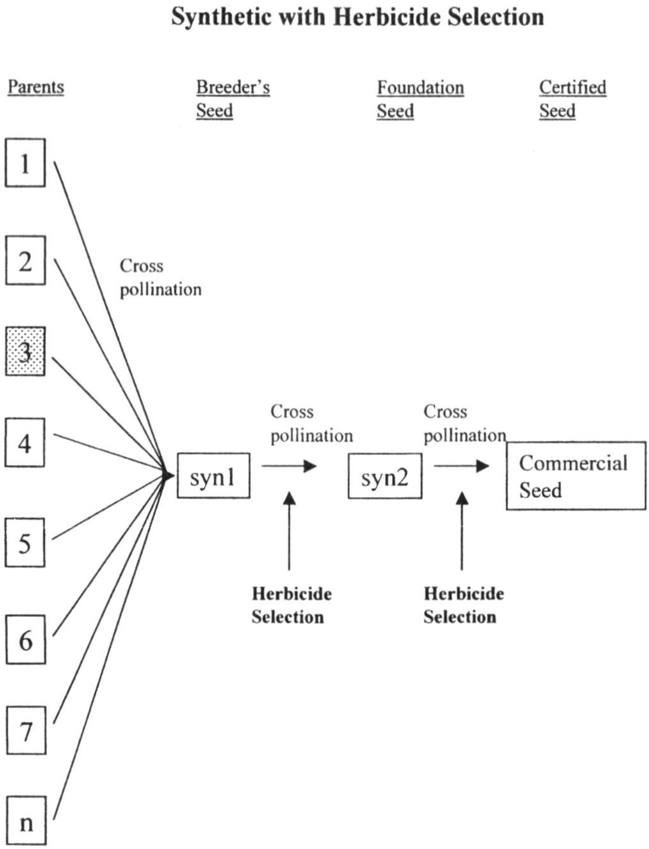

Fig. 9–5. Two seed production schemes using herbicide selection that might be used to deliver an alfalfa variety as a (A) conventional synthetic or (B) reduced generation synthetic from one transgenic parent (shaded).

would therefore contribute disproportionately to the composition of the synthetic variety. All plants carrying the transgene also would share a chromosome or chromosome segment containing the transgene and all linked genes both desirable and undesirable. The impact of the greater relationship among individuals in the population is unknown; however, if the variety is based on relatively few parents and a limited number of generations of seed increase, the impact would likely be minimal.

A final possibility would be to combine these two strategies and produce a double-cross hybrid or reduced generation synthetic using herbicide selection (Fig. 9–5B). If one of the parents was a transgenic expressing resistance to a particular herbicide and if herbicide selection was imposed on the Foundation seed production field, the majority of progeny would be hybrids between the transgenic parent (usually the male parent) and the other parents (usually the female parents).

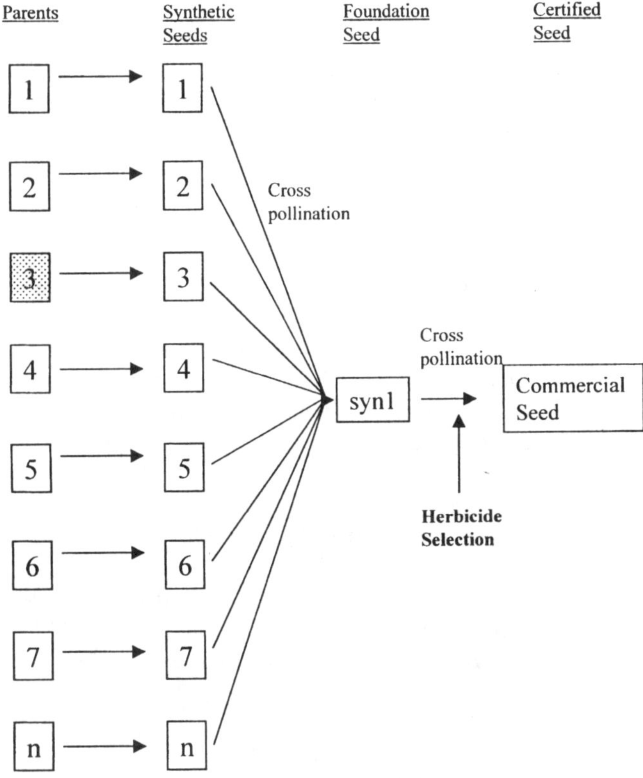


Fig. 9–5. Continued.

Table 9–4. Percentage of plants in a synthetic alfalfa variety carrying a single transgene linked to herbicide resistance increased through three seed generations with herbicide selection.†

| | Parents with transgene | | | | | |
|---|---|---|---|---|---|---|
| | Selection in the foundation field | | | Selection in the certified field | | |
| Generation | 100 | 50 | 10 | 100 | 50 | 10 |
| | % | | | | | |
| Breeder | 75.0 | 43.8 | 9.8 | 75.0 | 43.8 | 9.8 |
| Foundation | 84.8 | 79.5 | 75.8 | 84.8 | 79.5 | 75.8 |
| Certified | 81.9 | 75.9 | 71.7 | 89.5 | 85.9 | 85.3 |

† The model assumes a single gene insertion, chromosome segregation and autotetraploid inheritance.

## CONCLUSIONS

Somatic embryogenesis is fundamental to the development of genetically engineered alfalfa plants and to the development of synthetic seeds for plant propagation. Relatively few alfalfa plants have the ability to form somatic embryos and this limitation has greatly impeded the application of biotechnology in the improvement of this crop. On the other hand, because somatic embryogenesis is a heritable genetic trait, it has been possible using conventional breeding techniques to create populations of plants that have the ability to form somatic embryos as well as have good agronomic performance in terms of forage and seed yield, and stress and disease tolerances. The availability of this germplasm will ultimately speed the development of commercial cultivars from transgenic plants.

At present, selection for herbicide resistance appears to be an essential step during commercial seed increase to deliver a synthetic variety with a reasonably high proportion of the plants which contain the transgene. The best ways to accomplish this have not yet been determined experimentally.

## ACKNOWLEDGMENTS

Financial assistance for the research summarized in this review was provided by the Ontario Ministry of Agriculture, Food and Rural Affairs and the Natural Sciences and Engineering Research Council of Canada.

## REFERENCES

Austin, S., E.T. Bingham, D.E. Mathews, M.N. Shahan, J. Will, and R.R. Burgess. 1995. Production and field performance of transgenic alfalfa (*Medicago sativa* L.) expressing alpha-amylase and manganese-dependent lignin peroxidase. Euphytica 85:381–393.

Anandarajah, K., and B.D. McKersie. 1990. Enhanced vigor of dry somatic embryos of *Medicago sativa* L with increased sucrose. Plant Sci. 71:261–266.

Bingham, E.T., L.V. Hurley, D.M. Kaatz, and J.W. Saunders. 1975. Breeding alfalfa which regenerates from callus tissue in culture. Crop Sci. 15:719–721.

Borochov, A., M. A. Walker, E. J. Kendall, K. P. Pauls and B. D. McKersie. 1987. Effect of a freeze-thaw cycle on properties of microsomal membranes from wheat. Plant Physiol. 84:131-134.

Bowler, C., L. Slooten, S. Vandenbranden, R. de Rycke, J. Botterman, C. Sybesma, M. van Montagu and D. Inzé. 1991. Manganese superoxide dismutase can reduce cellular damage mediated by oxygen radicals in transgenic plants. EMBO J 10:1723–1732.

Bowler, C., M. van Montagu, and D. Inzé. 1992. Superoxide dismutase and stress tolerance. Annu. Rev. Plant Physiol. Plant Molec. Biol. 43:83–116.

Bowler, C., T. Alliotte, M. De Loose, M. Van Montagu, and D. Inzé. 1989. The induction of manganese superoxide dismutase in response to stress in *Nicotiana plumbaginifolia*. EMBO J. 8:31–38.

Bowler, C., W. van Camp, M. van Montagu, and D. Inzé. 1994. Superoxide dismutase in plants. Crit. Rev. Plant Sci. 13:199–218.

Bowley, S.R., G.A. Kielly, K. Anandarajah, , B.D. McKersie, and T. Senaratna. 1993. Field evaluation following two cycles of backcross transfer of somatic embryogenesis to commercial alfalfa germplasm. Can. J. Plant Sci. 73:131–137.

Bridger, G.M., W. Yang, D.E. Falk, and B.D. McKersie. 1994. Cold acclimation increases tolerance of activated oxygen in winter cereals. J. Plant Physiol. 144:235–240.

Brown, D.C.W., and A. Atanassov. 1985. Role of genetic background in somatic embryogenesis in *Medicago*. Plant Cell Tiss. Org. Cult. 4:111–122.

Crea, F., M. Bellucci, F. Damiani, and S. Arcioni. 1995. Genetic control of somatic embryogenesis in alfalfa (*Medicago sativa* L cv Adriana). Euphytica 81:151–155.

D'Halluin, K., J. Botterman, and W. de Greef. 1990. Engineering of herbicide-resistant alfalfa and evaluation under field conditions. Crop Sci. 30:866–871.

Desgagnes, R., S. Laberge, G. Allard, H. Khoudi, Y. Castonguay, J. Lapointe, R. Michaud, and L.P. Vezina. 1995. Genetic transformation of commercial breeding lines of alfalfa (*Medicago sativa*). Plant Cell Tiss. Org. Cult. 42:129–140.

Du S., L. Erickson, and S. R. Bowley. 1994. Effect of plant genotype on the transformation of cultivated alfalfa (*Medicago sativa*) by *Agrobacterium tumefaciens*. Plant Cell Rep. 13:330–334.

Foyer, C.H., P. Descourvieres, and K.J. Kunert. 1994. Protection against oxygen radicals: An important defence mechanism studied in transgenic plants. Plant Cell Environ. 17:507–523.

Fujii, J.A., D. Slade, J. Aguirre-Rascon, and K. Redenbaugh. 1992. Field planting of alfalfa artificial seeds. In Vitro Cell. Dev. Biol. 28P:73–80.

Fujii, J.A., D. Slade, and K. Redenbaugh. 1989. Maturation and greenhouse planting of alfalfa artificial seeds. In Vitro Cell. Dev. Biol. 25:1179–1182.

Gamborg, L.L., R.A. Miller, and K. Ojima. 1968. Nutrient requirements of suspension cultures of soybean root cells. Exp. Cell. Res. 50:151–158.

Gray, D.J., and A. Purohit, 1991. Somatic embryogenesis and development of synthetic seed technology. Crit. Rev. Plant Sci. 10:33–61.

Henry, Y., P. Vain, and J. DeBuyser. 1994. Genetic analysis of in vitro plant tissue culture responses and regeneration capacities. Euphytica 79:45–58.

Hernandez-Fernandez, M.M., and B.R. Christie, 1989. Inheritance of somatic embryogenesis in alfalfa (*Medicago sativa* L.). Genome 32:318–321.

Hetherington, P.R., H.L. Broughton, and B.D. McKersie. 1988. Ice-encasement injury to microsomal membranes from winter wheat crowns: II. Changes in membrane lipids during ice-encasement. Plant Physiol. 86:740–743.

Kendall, E.J., and B.D. McKersie. 1989. Free radical and freezing injury to cell membranes of winter wheat. Physiol. Plant. 76:86–94.

Kendall, E.J., B.D. McKersie, and R.H. Stinson. 1985. Phase properties of membranes after freezing injury in winter wheat. Can. J. Bot. 63:2274–2277.

Kielly, G.A., and S.R. Bowley. 1992. Genetic control of somatic embryogenesis in alfalfa. Genome 35:474–477.

Lai, F., and B.D. McKersie. 1994. Regulation of storage protein synthesis by nitrogen and sulphur nutrients in alfalfa (*Medicago sativa* L.) somatic embryos. Plant Sci. 103:209–221.

Lai, F.M., C.G. Lecouteux, and B.D. McKersie. 1995. Germination of alfalfa (*Medicago sativa* L.) seeds and desiccated somatic embryos:1. Mobilization of storage reserves. J. Plant Physiol. 145:507–513.

McKersie, B.D., and S.R. Bowley. 1993. Synthetic seeds in alfalfa. p. 231–255. *In* K. Redenbaugh (ed.) Synseeds applications of synthetic seeds to crop improvement. CRC Press, Boca Raton, FL.

McKersie, B.D., S.R. Bowley, E. Harjanto, and O. Leprince. 1996. Water-deficit tolerance and field performance of transgenic alfalfa overexpressing superoxide dismutase. Plant Physiol. 111:1177–1181.

McKersie, B.D., and D.C.W. Brown. 1996. Somatic embryogenesis and artificial seeds in forage legumes. Seed Sci. Res. 6:109–126.

McKersie, B.D., Y.R. Chen, M. deBeus, S.R. Bowley, C. Bowler, D. Inzé, K. D'Halluin, and J. Botterman. 1993. Superoxide dismutase enhances tolerance of freezing stress in transgenic alfalfa (*Medicago sativa* L.). Plant Physiol. 103:1155–1163.

McKersie, B.D., T. Senaratna, S.R. Bowley, D.C.W. Brown, J.E. Krochko, and J.D.Bewley. 1989. Application of artificial seed technology in the production of hybrid alfalfa (*Medicago sativa*). In Vitro Cell. Dev. Biol. 25:1183–1188.

Meijer, E.G.M., and D.C.W. Brown. 1985. Screening of diploid *Medicago sativa* germplasm for somatic embryogenesis. Plant Cell Rep. 4:285–288.

Micallef, M.C., S. Austin, and E.T. Bingham. 1995. Improvement of transgenic alfalfa by backcrossing. In Vitro Cell. Dev. Biol. 31P:187–192.

Murashige, T., and F. Skoog. 1962. A revised medium for rapid growth and bioassays with tobacco cultures. Physiol. Plant. 15:473–479.

Narvaez-Vasquez, J., M.L. Orozco-Cardenas, and C.A. Ryan. 1992. Differential expression of a chimeric CaMV-tomato proteinase inhibitor I gene in leaves of transformed nightshade, tobacco and alfalfa plants. Plant Molec. Biol. 20:1149–1157.

Redenbaugh, K. 1993. Synseeds: Applications of synthetic seeds to crop improvement. CRC Press, Boca Raton.

Reisch, B., and E.T. Bingham. 1980. The genetic control of bud formation from callus cultures of diploid alfalfa. Plant Sci. Lett. 20:71–77.

Samac, D.A. 1995. Strain specificity in transformation of alfalfa by *Agrobacterium tumefaciens*. Plant Cell Tiss. Org. Cult. 43:271–277.

Saunders, J.W., and E.T. Bingham. 1972. Production of alfalfa plants from callus tissue. Crop Sci. 12:804–808.

Scandalias, J.G. 1990. Response of plant antioxidant defense genes to environmental stress. Adv. Genet. 28:1–41.

Scandalias, J.G. 1993. Oxygen stress and superoxide dismutase. Plant Physiol. 101:7–12.

Schenk, B.U., and A.C. Hildebrandt. 1972. Medium and techniques for induction and growth of monocotyledonous and dicotyledonous plant cell culture. Can. J. Bot. 50:199–204.

Schubert, B.K. 1994. Glutathione, glutathione reductase and freezing stress in alfalfa (*Medicago sativa* L.). M.S. thesis. Univ. of Guelph, Guelph, ON.

Seitz-Kris, M.H., and E.T. Bingham. 1988. Interactions of highly regenerative genotypes of alfalfa (*Medicago sativa* L.) and tissue culture protocols. In Vitro Cell. Dev. Biol. 24:1046–1052.

Senaratna, T., B.D. McKersie, and S.R. Bowley. 1989. Desiccation tolerance of alfalfa (*Medicago sativa* L.) somatic embryos. Influence of abscisic acid, stress pretreatments and drying rates. Plant Sci. 65:253–259.

Senaratna, T., B.D. McKersie, and S.R. Bowley. 1990. Artificial seeds of alfalfa (*Medicago sativa* L.) Induction of desiccation tolerance in somatic embryos. In Vitro Cell. Dev. Biol. 16:85–90.

Senaratna, T., B.D. McKersie, and R.H. Stinson. 1985. Simulation of dehydration injury to membranes from soybean axes by free radicals. Plant Physiol. 77:472–474.

Shahin, E.A., A. Spielmann, K. Sukhapinda, R.B. Simpson, and M. Yashar. 1986. Transformation of cultivated alfalfa using disarmed *Agrobacterium tumefaciens*. Crop Sci. 26:1235–1239.

Spano, L., D. Mariotti, M. Pezzotti, F. Damiani, and S. Arcioni. 1987. Hairy root transformation in alfalfa (*Medicago sativa* L.). Theor. Appl. Genet. 73:523–530.

Thomas, J.C., C.C. Wasmann, C. Echt, R.L. Dunn, H.J. Bohnert, and T.J. McCoy. 1994. Introduction and expression of an insect proteinase inhibitor in alfalfa (*Medicago sativa* L). Plant Cell Rep. 14:31–36.

Wenzel, C.L., and D.C.W. Brown. 1991. Histological events leading to somatic embryo formation in cultured petioles of alfalfa. In Vitro Cell. Dev. Biol. 27P:190–196.

Yamada, T., and K. Okumura. 1997. Germplasm conservation. p. 61–89. *In* B.D. McKersie and D.C.W. Brown (ed.) Biotechnology and the improvement of forage legumes. CAB Int., Wallingford, England.